Establishment of a Geologic Framework for Paleoanthropology

Edited by

Léo F. Laporte
Earth Sciences
University of California
Santa Cruz, California 95064

THE GEOLOGICAL SOCIETY OF AMERICA · 1888 ·

SPECIAL PAPER

242

1990

Published by The Geological Society of America, Inc.
3300 Penrose Place, P.O. Box 9140, Boulder, Colorado 80301

GSA Books Science Editor Richard A. Hoppin

Printed in U.S.A.

Library of Congress Cataloging-in-Publication Data
Establishment of a geologic framework for paleoanthropology / edited
by Léo F. Laporte.
p. cm. — (Special paper / Geological Society of America ;
242)
"A contribution of the History of Geology Division of the
Geological Society of America to the centennial celebration of the
Society"—Pref.
Includes bibliographical references.
ISBN 0-8137-2242-X
1. Fossil man. 2. Human evolution. 3. Anthropology, Prehistoric.
4. Geology, Stratigraphic. I. Laporte, Léo F. II. Geological
Society of America. History of Geology Division. III. Series:
Special papers (Geological Society of America) ; 242.
GN281.4E83 1990
573.2—dc20 89-25989
CIP

Cover photo: Site of the hominid skull, "1470," attributed to the pre-*Homo erectus* species, *Homo habilis,* at Lake Turkana, northern Kenya. Sediments are tuffaceous sands deposited in a distributary channel and lie about the Pliocene-Pleistocene boundary, having been dated at some 2 Ma. Although a surface find, matrix attached to skull fragments indicates its probable stratigraphic position, now marked by the concrete plinth. Photo by L. F. Laporte.

10 9 8 7 6 5 4 3 2

Contents

Preface

This Special Paper is a contribution of the History of Geology Division of the Geological Society of America to the centennial celebration of the Society. Several years ago, when the Division was considering how best to celebrate the centennial, we recalled the words of the American paleontologist, George Gaylord Simpson, who asserted that the "most interesting thing about the world is life, and the most interesting thing about life is evolution." Taking Simpson's lead, we further concluded that the most interesting thing about evolution—to a human, that is—is the origin and development of the human lineage.

One of the great contributions to human knowledge has been deciphering the fossil history of our own lineage within the evolutionary tree of animate nature on this planet. And as is obvious to all students of human evolution, the geological sciences have made crucial contributions to this knowledge by establishing a temporal framework—in terms of relative and absolute time—within which human origins and subsequent evolution can be understood. In addition to this temporal framework, geology has provided information about the environments in which that evolution occurred.

Therefore, during our centennial celebration, it seemed fitting to pause and reflect on a few of these critical steps in our understanding of the geological framework of early human origins. Although representing just a small part of the geological knowledge that has illuminated our early prehistory, the papers in this volume do suggest the nature of the contribution.

Donald Grayson traces the history of the eventual recognition that there was, indeed, "deep human antiquity," contrary to both Buffon and Cuvier who, in the late eighteenth and early nineteenth centuries, claimed that humans did not make their appearance until the "modern" world existed. As Grayson explains, careful geological exploration of Brixham Cave in 1858 showed that flint artifacts were directly associated with the bones of extinct elephants, rhinoceroses, cave bears, and hyenas, thereby demonstrating that the human species was quite ancient. This definitive association of stone tools with extinct mammals was finally accepted as valid in 1859, independent of the publication in the same year of Darwin's *On the Origin of Species.* But this conjunction of Darwinian "descent with modification" with proof of human antiquity had, of course, enormous implications for human origins.

William Farrand next considers the development of thought with respect to that interval of geologic time during which our human ancestors first appeared and subsequently evolved. It is not enough to know that our lineage is "ancient"; we also want to reliably place various aspects of our history within a finely resolved time sequence. Until that sequence, embracing "Quaternary," "Holocene," and "Pleistocene," is firmly established, further details of climate, environment, biota, and absolute ages cannot be established. Farrand sorts out the conceptions and misconceptions applied to these last two million years of geologic time, starting with Arduino and Mantell and ranging through the later work of Desnoyers and Morlot as well as Lyell and Forbes, coming up to the debates of the present day.

Richard Hay presents a case history of the geological interpretation of early human habitats in East Africa. The sedimentary strata of Olduvai Gorge have produced a superb sequence of artifacts and hominid remains, but rapid facies changes of the sediments within short distances made initial correlation of the individual strata difficult. Hay emphasizes that this difficulty was compounded by erroneous interpretations of Pleistocene climate that were first used for correlation. Despite a half-century of study, it wasn't until the 1960s that a detailed stratigraphy was established. Hay then summarizes the results (and debates) of paleoenvironmental analyses, bed by bed, within Olduvai Gorge strata. What makes Hay's discussion especially interesting is that much of what we've come to understand about the geology of Olduvai Gorge has been due to his own research efforts.

B. Theunissen, J. de Vos, and P. Y. Sondaar review the efforts to determine the age of so-called "Java Man," the first hominid fossil to be accepted as evidence for human evolution. The central figure in this story is, of course, Eugene Dubois, whose unflagging energy and enthusiasm resulted in the original discoveries at the turn of this century. As the authors indicate, confusion of lithostratigraphy and biostratigraphy, compounded by assumptions of primitive versus advanced morphology, has obscured the dating of these fossils, which has been disputed since their initial discovery and continues right up to the present. More recent absolute age dating has also had its own shortcomings. The authors conclude their paper with a summary of the results of their own recent work, and they propose a revised chronology for the Javanese hominid-bearing deposits.

Vance Haynes brings the discussion to the New World and focuses on the contributions of two twentieth century geologists, Kirk Bryan and Ernest Antevs, working with archaeologists in the attempt to establish a Paleoindian chronology. Antev's chronology was based on the climatic interpretation of varved sequences in the Great Basin, following techniques developed in Scandinavia, whereas Bryan used alluvial surficial geology to establish climatic sequences in the western United States. Both men applied their climatic chronologies to dating the Folsom and Clovis cultures and, although they did not initially agree, they demonstrated the potential usefulness of their techniques. Haynes discusses the subsequent development of "geologic-climatic dating," the use of radiocarbon dating in establishing absolute ages, and some of the larger questions that deal with establishing time of human entry into North America.

Zhen and Ho recapitulate the history of the efforts to determine the geologic age of the *Homo erectus*-bearing cave deposits at Zhoukoudian ("Choukoutien"), China. Using associated fauna and sedimentary facies, initial efforts at dating these deposits in the earlier decades of this century suggested a range of ages, from late Pliocene to middle Pleistocene. Broader theories of climate and morphological evolution tended to confound the age interpretations. Not until more recent use of a variety of absolute-age dating techniques has it been possible to narrow the age of the hominid-bearing deposits from some 500,000 years, near the base of the deposits, to about 200,000 years at their top. Zhen and Ho interpret these new dates as to what they might indicate about the evolution of the Asian lineage of *H. erectus.*

Clearly, in the century since the founding of the Geological Society of America in 1888, we have come a long way in understanding the timing and environmental setting of early human history. And as the papers in this volume indicate, geology has played a significant role in the study of human origins and evolution. Equally important, however, is our understanding of how the science of geology proceeds, and these papers demonstrate that the forward progress of scientific knowledge is anything but unidirectional and inevitable. The practice of science, like other human endeavors, may depend as much on initial assumptions, vagaries of human personality, state of technology, and serendipity of discovery, as on right reason and unburdened intelligence.

* * *

The History of Geology Division dedicates this volume to the late Claude C. Albritton, Jr., of Southern Methodist University, who made so many valuable contributions to the study of the philosophy and history of geology. In fact, it was 25 years ago, during the commemoration of the 75th anniversary of the Geological Society of America, that Albritton organized a symposium, as well as editing the resultant volume, on the *Fabric of Geology,* which examined some of the epistemological foundations of our science. In his 1980 book, *The Abyss of Time,* Albritton remarked that "the concept of the vast extension of our world in time . . . has surely been no less revolutionary than the concept of vast stellar space. However, . . . the idea of geological time, as often as not, has been greeted with indifference, skepticism, or downright hostility."

Leó F. Laporte, Chairman,
History of Geology Division, 1988.

Geological Society of America
Special Paper 242
1990

The provision of time depth for paleoanthropology

Donald K. Grayson
Department of Anthropology and Burke Memorial Museum, University of Washington, Seattle, Washington 98195

INTRODUCTION

"There is something dreary in the indefinite lengthening of a savage and blood-stained past" wrote Julia Wedgwood for Macmillan's Magazine in April, 1863 (p. 486). These words were a mild reaction, perhaps fitting for Darwin's niece, to the recent recognition that humankind had been on earth for a remarkably long period of time. If dreary were all there were to it, then the establishment of a deep human antiquity in 1859 would not represent the most significant contribution made by archaeological data to our understanding of our own place in the scheme of things. But the implications of this recognition went almost indescribably further than dreary, and few other facts provided by ancient human remains have cut deeper.

The establishment of a proper order of human antiquity is often presented in terms of the absolute chronology involved. The events culminating in that establishment in 1859 are seen as momentous primarily because they so powerfully refuted the date of roughly 6,000 years provided for the human creation by theologians and theologically inspired scientists. But while the refutation did occur, relative chronology was by far the more important issue, since the new discovery refuted as well, although perhaps not so clearly, the twin beliefs that the history of the Earth was the history of our planet as it was being prepared for our arrival by a wise and caring Creator, and that the human species was central to the meaning of it all.

THE DISCOVERY OF STONE TOOLS

The deeper annals of our past are written primarily in stone tools; unfortunately, the bones of the makers of those tools, to continue the analogy, are more like footnotes to a sparsely referenced text. As a result, it was the discovery of such tools, not of human bones, in undeniably ancient contexts that led to the massive restructuring of our past during the last half of the nineteenth century.

It also had to be discovered that stone tools really are stone tools, but that discovery had taken place much earlier. During the late 1600s, the concept of fossils as all things distinctively shaped, stony, and from in or on the ground had been shattered, revealing classes of objects, including stone tools, that were now to be accounted for in different ways (Rudwick, 1972; Grayson, 1983: additional details on much, though not all, of what follows in this chapter can be found in the latter reference). As soon as stone tools were recognized as such, it was recognized as well that European examples must have been relatively ancient, since no one would use stone to cut, chop, and pierce if they had the ability to use metal implements to serve those functions. By the end of the eighteenth century, it was widely agreed not only that stone had preceded metal for the manufacture of tools, but that there had been a chronological sequence in the kinds of metals that had been employed, with brass or bronze having come first, iron later. Even though European stone tool use was seen as having been early, and although the peoples involved were shrouded by time, there was no reason to think that the users of those tools predated events described in the Bible. Events in paleontology and geology ultimately lifted this shroud.

BUFFON'S INTERPRETIVE FRAMEWORK

During the late 1600s and early 1700s, there was widespread agreement that the Earth had been created about 6,000 years ago, a date that had been reached by tallying the biblical genealogies and then combining those tallies with analyses of other historical documents and with various astronomical calculations. Archbishop James Ussher's date of 4004 B.C. for the Creation may be the most famous of those dates, but it was but one of many. All seemed to indicate that the Creation had occurred roughly 6,000 years ago.

The 6,000-year date for the creation of the Earth actually applied to the creation of human beings, since the heart of the chronology was based on genealogies that began, of course, with Adam. The extension of this date to all of earth history followed from a literal reading of the word "day" in the verses of Genesis dealing with events before Adam. It also followed, however, from the widely held assumption that the Earth had been created for our species by a caring, if occasionally wrathful, Creator, and that an Earth without people was an Earth without purpose (see, e.g., Porter, 1977).

There was, then, a flexibility inherent in the biblical chronol-

Grayson, D. K., 1990, The provision of time depth for paleoanthropology, *in* Laporte, L. F., ed., 1990, Establishment of a geologic framework for paleoanthropology: Boulder, Colorado, Geological Society of America Special Paper 242.

ogy, since doubts that might arise concerning the meaning of the pre-Adam days could allow the pre-human epochs of earth history to become longer, while the beginnings of human history remained fixed at roughly 6,000 years ago. This flexibility was most directly exploited in France. The French Enlightenment saw the Creator recede into the background as a continual influence in determining the nature and direction of earth history. For some French writers, God drifted out of the picture entirely, but more commonly, the Creator simply remained offstage, leaving an Earth that marched along on its own. In this world view, the scriptural account of earth history played little or no role; the Earth became a body of great antiquity whose past had been marked by continual and dynamic change (see, for instance, Burchfield, 1975; Davies, 1969; Gay, 1966; Porter, 1977; Laudan, 1987).

Georges Buffon, perhaps the finest naturalist of his century, most successfully captured and popularized this dynamic view of nature and of earth history. Buffon felt himself to be a disciple of Newton, searching for the laws that had governed the development of our planet, and allowed Scripture to play a determining role only as it related to our own species. In his *Epochs of Nature* (1778), he organized the deeper history of the Earth in terms of seven major epochs, thus preserving the form, though few of the details, of the biblical creation story. Assuming that the Earth had formed from an incandescent mass, he calculated that our planet had been in existence for at least 75,000 years. People did not appear on this dynamic Earth until the end of his sixth epoch; the seventh, and final, epoch saw the rise of human power as a force in modifying the surface of the Earth.

If the biblical chronology did not apply to earth history, did it apply to anything? Here, Buffon capitalized on the flexibility inherent in that chronology, unhooking it from earth history and applying it to human history alone: "since the end of the works of God, that is to say since the creation of man, only six or eight thousand years have passed, because the genealogies of mankind since Adam indicate no more" (1778, p. 35).

The biblical chronology, then, applied to people and to people alone. But there was far more to Buffon's position than that. "Since the end of the works of God" may have been a stock phrase, but it also encapsulated his view that we had not appeared on Earth until the Earth had been fully prepared for us, until it had become modern in form. The Supreme Being, Buffon (1778, p. 189–190) argued,

> provided sufficient time for the Earth . . . to arrive, finally, at that state of calm and tranquillity in which man could be the intelligent witness, the peaceful admirer, of the grand spectacle of Nature and of the marvels of creation. Thus we are persuaded, independently of the authority of the Sacred Books, that man was created last and that he did not come to take the sceptre of the Earth until it was worthy of his rule.

Buffon thus combined three separate themes into his powerful synthesis of earth history: an ancient planet, the origin of the human species some 6,000 years ago, and the correlation between the appearance of people and the development of a physically modern Earth. His *Epochs of Nature* was the first substantial work by a renowned natural historian to present this combination of ideas. The combination was powerful, joining an older belief that earth history had been directional with a growing belief that it had also been longer than Scripture seemed to allow, while retaining the accuracy of Scripture on what was, after all, the most important point: the early history of human beings. Humankind had not appeared until the rest of the Creator's work had been accomplished, and the Earth had become modern in form, some 6,000 years ago.

At the beginning of the eighteenth century, I have noted, it was widely accepted that an Earth without people was an Earth without purpose. By the end of the century, the first few days of Genesis had become allegory for most, and earth history had become tremendously long, with the human species appearing only toward the end of that history. Now, and especially in England, the history of the Earth became the history of our planet as it developed into a fit habitat for humankind, a history that had been superintended, at one degree of remove or another, by a Creator who cared for all of his beings (see, for instance, Rudwick, 1972). An Earth without people was no longer an Earth without purpose, and this view was still firmly in line with Scripture.

CUVIER AND THE DIRECT MEASUREMENT OF TIME

Given this set of beliefs, the discovery of human remains in deposits that could not be considered geologically modern would have tremendous implications. Since the human appearance was seen to have postdated the onset of modern geologic conditions, such a discovery would call into question the accuracy of the biblical genealogies, and thus of Revelation itself. Since the human species was God's last creative work, as Scripture made clear, such a discovery would conflict with God's Words in this way as well. By showing that people had appeared on Earth before the world had been fully prepared for their arrival, such a discovery would shed doubt on the belief that the development of the Earth had been superintended by a caring Creator.

Even for those unconcerned with the direct reconciliation of the facts of geology with Scripture, these beliefs formed a significant background for the reception of the relics of possibly ancient peoples. Even without this background, it was widely believed that each kind of organism had been created in one place from which it had subsequently spread, and that each kind had been adapted from the beginning to a certain set of environmental conditions (see, for instance, Ospovat, 1981). This view was applied to our species as well, with western Asia routinely considered as a homeland, as both Scripture and the records of the most ancient civilizations suggested. In such a view, the creation of humankind prior to the onset of modern environmental conditions simply did not seem possible.

At the time Buffon wrote, the belief in human recency was

secure from being questioned on geological grounds. Human remains were clearly absent from such undoubtedly ancient strata as those that provided coal, while there were no known attributes of the most superficial strata of the earth that marked the termination of pre-modern, or the onset of modern, times with any clarity. Indeed, given this situation, the very presence of human bones or artifacts implied the recency of the deposits in which they had been found (as, for instance, in Esper, 1774). This situation changed dramatically at the end of the eighteenth century, as a set of geographically widespread time markers, linked with what were seen as major and geologically recent events in earth history, were defined for the superficial strata. These markers allowed the uppermost strata to be placed securely in the general framework of earth history, whether or not they contained human remains. Since they were initially also linked with the biblical chronology, the markers provided direct empirical access to both chronologies, relative and absolute, that had been accepted for human history on Earth.

This means of reading modernity from the geologic record was developed by the brilliant French comparative anatomist Georges Cuvier. Cuvier's successful demonstration of the reality of extinction was built on the detailed morphological examination of mammals so large that their future discovery on the hoof could not be reasonably entertained (Rudwick, 1972; Grayson, 1984). A number of the animals Cuvier examined, including mastodon and mammoth, were, we now know, Pleistocene in age; they had, accordingly, been discovered in the most superficial strata of the earth, and were the most recent of the extinct organisms he had described. Finding the kinds of slowly operating geologic processes seen in action today to be insufficient to account for the major changes he saw in the geologic record, Cuvier argued that these animals had become extinct as a result of some rapid change, some revolution as he called it, that had affected the surface of the Earth. "If there is anything that is established in geology," he wrote in his *Preliminary Discourse* of 1812, "it is that the surface of our globe was the victim of a great and sudden revolution that can date to no more than five or six thousand years ago" (1812, p. 110), a revolution that saw low-lying areas inundated and the former sea bed exposed. This rapid, but transient, incursion of the sea caused the extinction of those mammals caught in its way. Recolonization of flooded areas then occurred from regions not so affected, but such modern animals as elephants and rhinoceroses could not have descended from the extinct European forms because those species had been totally annihilated.

Cuvier stressed flooding because the remains of the extinct mammals he had described had often been found associated with a set of superficial gravels, gravels whose form and position are now explained in terms of events associated with glaciation. Cuvier followed J.-A. de Luc (1779) in attributing these gravels to flooding—the only mechanism then known that could have accounted for them—and was influenced by de Luc as well in dating the flooding event to roughly 6,000 years ago (Rudwick, 1972).

For Cuvier, then, the superficial gravels and the extinction of such mammals as the mammoth and hippopotamus in Europe became tightly linked to one another chronologically, since both had been caused by the same event some 5,000 to 6,000 years ago. It was also obvious to Cuvier that the areas affected by the flooding took on their modern appearance only after the last retreat of the sea, and that it was also at that time that the nature of the surface of the earth in those regions began to be determined solely by operation of those processes that can be seen in action today.

Cuvier's arguments provided just those geologic markers of modernity that Buffon's synthesis lacked: the superficial gravels and extinct mammals, dated to about 6,000 years ago, represented the end of the pre-modern Earth in those vast regions of the world where they were found. If people had existed on a pre-modern Earth, their remains would be found associated with these time markers. Cuvier, of course, was fully aware of this fact, and he carefully observed that the absence of human bones in deposits with the remains of the extinct quadrupeds implied the absence of people in these areas at the time the deposits accumulated. "There are no fossil human bones," Cuvier (1812, p. 82) emphatically observed. He did not deny that people had existed somewhere on Earth prior to the revolution that had caused the extinctions and deposited the gravels. Instead, he argued that they had not existed prior to this time in the geologically known portions of Europe, Asia, and America. Those regions they had entered, like all other mammals, after the waters had retreated.

To a large degree, Cuvier's conclusion that the most recent episode of extinctions had occurred before the arrival of human beings could have been inductively derived from the fossil record itself. In 1812 there was no compelling evidence for any association between human relics and either the extinct mammals or the diluvium. While such evidence was largely lacking because the time markers did not become such until Cuvier made them so, it is also true that Esper (1774) had described what could have been just such an association at Gailenreuth Cave in Germany. Cuvier was fully familiar with Esper's work, and had he any desire to make the argument for a deep human antiquity, he could have used Gailenreuth Cave to do so. In addition, while there were no truly convincing cases in 1812, he continued to make the same argument until his death in 1832, by which time a number of apparent associations between human remains and those of extinct mammals had been published and widely discussed (see below).

Instead, Cuvier's position on human recency was largely theoretical in origin. Viewing a species as a group of organisms of common descent and functionally integrated to meet the demands of the environment in which they existed, Cuvier saw each species as having originated in any given area only when the environments to which they were adapted had appeared. His reconstruction of extinct vertebrates, his knowledge of fossil invertebrates, and his use of stratigraphic position to place these animals in an ordered scheme of earth history allowed him to show that the history of life on Earth had not only been directional, but also that the nature of this direction had apparently been determined by the nature of the changing environments of the Earth itself (Burkhardt, 1977). Cuvier's position on the recency of the human species in all areas affected by the last revolution was fully consistent with this view.

But while these positions were consistent, Cuvier's careful arguments against the association between human relics and the remains of the extinct mammals stemmed more directly from his concerns for documenting the reality of extinction itself, and in particular from his opposition to Jean-Baptiste Lamarck's transformationist hypotheses. To Lamarck, the general design and balance of nature seemed to deny the possibility of extinction for most kinds of animals. Animals known only from their ancient remains represented forms that had simply not yet been discovered alive, or the ancestors of transformed descendants. Extinction, he argued, could only affect simple forms of life directly exposed to the harshness of nature, or large, slow-breeding animals with little defense against human predation (Lamarck, 1801, 1809). These latter, of course, were just the kind Cuvier had described. In arguing that the species he had reconstructed had been totally destroyed, Cuvier was removing a potential set of ancestors for a number of modern mammals. Since Lamarck was willing to concede this point, by stressing the absence of human remains with the bones of the animals themselves, Cuvier was removing the only cause of extinction for these creatures that Lamarck had allowed. These were extinct forms, people could not have caused their extinction because people simply had not been there, and some other, natural cause must have been involved (see the excellent discussions in Rudwick, 1972; Burkhardt, 1977; Appel, 1987).

Yet another reason, however, may have been uppermost in Cuvier's mind. The most powerful known cause of morphological variability within species was domestication, and Lamarck (1809) had stressed the power of that source to support the possibility of transformation. Now, Cuvier had to demonstrate that this potent agent could not account for the characters that distinguished his extinct mammals from similar modern species. Accordingly, Cuvier (1812) inserted his argument that "there are no fossil human bones" immediately after his discussion of variability within species, including variability caused by domestication. And, he began his argument against the existence of human fossils by observing that the lack of human bones among the remains of extinct mammals provides "more proof that the fossil races were not varieties [of modern species] since they could not have been subject to the influence of man" (1812, p. 82). Without a human presence, Cuvier's extinct mammals could not owe their distinctive form to human modification.

Cuvier's stance on human antiquity dictated the nature and course of research in this area for the next 50 years. He had taken Buffon's synthesis concerning early human history, in which the human lineage was seen as having appeared only after the Earth had become modern in form and as having appeared some 6,000 years ago, and had provided the empirical correlates whereby these chronological arguments could be directly tested. Further, he had shown precisely how the test could be done, and had shown as well the kind of information that appeared to be needed to make that test.

Cuvier's work had tremendous impact in Great Britain, where it was quickly converted into a piece of natural theology, a process that began even before Cuvier's *Preliminary Discourse* was published. In the third volume of his *Organic Remains of a Former World* (1811), James Parkinson (for whom the disease is named) depended heavily on Cuvier's published accounts of ancient vertebrates, and found the relationship between what was then known about earth and life history and the Mosaic account of that history so close "that the Mosaic account is thereby confirmed in every respect, except as to the age of the world" (1811, p. 451). Even this one disjunction disappeared as soon as the Mosaic day was taken as allegory. For Parkinson, the superficial gravels became evidence for the Deluge; the lack of ancient human relics, proof that our species was God's last creative work, "the last and highest work appearing to be *man*" (1811, p. 455). This equation, of course, became a problem, for if the gravels represented the Deluge, then people must have existed at the time, but if they had existed, they would have done so in a world not yet modern in form. "Organic Remains of a *Former* world" could not include human remains. This was a real dilemma, one which Parkinson explicitly recognized, but not one he dealt with.

Parkinson's arguments laid the groundwork for the British approach to human antiquity during the early nineteenth century. He had taken the paleontological evidence for the history of life on Earth, much of it provided by Cuvier, shown that it meshed with the Mosaic account, and built human history into that picture, with our species appearing last in a progressive sequence of creations. While Parkinson laid the groundwork, and Robert Jameson helped solidify it with his introduction to the 1813 English translation of Cuvier's *Preliminary Discourse* (Rudwick, 1972), it was William Buckland who completed the process.

Buckland's 1819 inaugural lecture at Oxford, meant in large part to stress the worldly importance of geology (Rupke, 1983), emphasized that geology showed the Mosaic account of Creation and subsequent events to be correct in all major areas. Most importantly, these areas included "the two great points . . . of the low antiquity of the human race, and the universality of a recent deluge" (1820, p. 24). His *Reliquiae Diluvianae* (1823), written explicitly to deal with the second of these two great points, appeared three years later. While his aims and concerns were much the same as Parkinson's, Buckland was a superb scientist, a superb natural theologian, and a superb writer. Using evidence that was later to be accounted for by glacial theory in a very different way, his careful, tight arguments attributed the deposition of what he called "diluvium," "those extensive and general deposits of superficial loam and gravel" (1823, p. 2), to the Deluge. The same Deluge that deposited the diluvium also accounted for the extinction of the associated mammals, "all destroyed together, by the waters of the same inundation which produced the deposits of loam and gravel in which they are embedded" (1823, p. 184).

Buckland thus produced a universal deluge where Cuvier had posited something much less. Just as Cuvier and Parkinson before him, Buckland argued that no human remains had been found securely in or beneath the diluvium, or in any setting that suggested contemporaneity with the now-extinct mammals. To argue otherwise would have placed people in a non-modern world, and would have opened the door to the possibility of vastly early peoples. Just like Parkinson, Buckland was acutely aware of the dilemma this created: how could there be no evidence for pre-Deluge peoples when the very reason for that flood was "to put an end to all people" (Genesis 6:13), with the exception of Noah and his family? Rather than dodging the issue, as Parkinson had done, Buckland tackled it head-on. The remains of antediluvian peoples were to be found in central or southern Asia, the cradle of the human race. This was also an area in which the extinct mammals of the diluvium were as yet unknown: Buckland reported only horse and deer, fully modern forms, from this part of the world. All this formed a fully consistent picture. People were recent in Europe but more ancient in Asia, just as Scripture suggested. Nowhere were they known from an area or a time that was not yet modern in form.

The equation between diluvium and Deluge was short lived. By 1830, mainstream British geologists had recognized that the evidence was inconsistent with such an equation, and the idea was abandoned (e.g., Sedgwick, 1834a, 1834b; see also Gillispie, 1959; Rupke, 1983). Parkinson's dilemma, the lack of human remains in or beneath the diluvium, had been resolved. The general British abandonment of this equation completely altered the chronological and environmental implications of the event with regard to human antiquity. If the diluvium had not been deposited by the flood, it could not be linked with the biblical chronology. The loss of this absolute time marker meant the loss of any immediately convincing scientific access to that chronology derived from the lack of association between human remains and either the extinct mammals or the diluvium. As a result, explicit references to dates derived from the Mosaic chronology become difficult to find after about 1830, although they were not uncommon before that time. This is not to say that absolute chronology had become unimportant. As Darwin wrote to his sister Caroline in 1837, "As far as I known everyone has as yet thought that the six thousand odd years has been the right period" (Burkhardt and Smith, 1986, p. 8), and the demise of that chronology was a source of major concern after a deep human antiquity was established in 1859. Instead, once the diluvium and the extinctions, on the one hand, and the Deluge, on the other, were decoupled, direct scientific access to absolute chronology was lost, and explicit concern with the biblical chronology receded far into the background.

The decoupling had another effect as well, for if the diluvium had not resulted from the Deluge, it clearly had to be far older: the fauna alone now seemed to suggest that. The increased perceived age of the diluvium and its mammals thus made an association between either of those markers and human remains both far less likely and far more dangerous. So did the fact that it was now clearer than ever that the Earth had not become modern in form until after the extinctions and until after deposition of the superficial gravels. These were clearly, as Adam Sedgwick said, "the remnants of a former world" (1834b, p. 313).

Thus, during the first few decades of the nineteenth century, Buffon's popular vision of early human history had been transformed both in Britain and on the continent into a set of very explicit expectations about when human relics should begin to appear in the geologic record, and as to the nature of the fauna that should accompany that record. Although, for reasons that were primarily theological in origin, the development of these expectations differed in Britain and on the continent, the results were the same: people did not appear on Earth until after the extinctions had occurred. Were this shown to be false, major consequences for critical western beliefs would follow.

PAUL TOURNAL: A BRIEF CASE STUDY

Even though the discovery of human remains with those of the extinct mammals would be full of unwelcome meaning, many such discoveries were reported between 1812, when Cuvier's synthesis first appeared, and 1858, when the final scene opened. Most of the sites on the list were from continental Europe, since the question of human antiquity here was not as thoroughly charged with theological overtones as it was in Britain, where external evidence relating to the factual nature of the events discussed in the Bible was of far more importance.

The majority of the European sites argued to contain the commingled remains of people and of extinct mammals were reported by a small number of extremely active people. These individuals had in common the fact that they were not members of the scientific elite. Most, in fact, were provincial, living far from the centers of science in their respective countries. Those who did not share those attributes yet nonetheless found evidence suggesting contemporaneity did not press the issue; the Parisian geologist Ami Boué, a founding member of the Geological Society of France, provides an excellent example (e.g., Boué, 1825, 1829, 1830). The attempt to make the case came primarily from the provinces, while the elite of geological science, located in the geographic heart of the scientific establishment, rejected that attempt.

The fact that many of these people were provincial does not mean that they were unaware of the significance of what they had found, although not all had the conceptual tools needed to deal with their discoveries in a convincing way. For instance, Marcel de Serres—lawyer, professor of geology and paleontology at the Faculty of Sciences in Montpellier, and prolific author—excavated a number of caverns in southern France that seemed to provide evidence of contemporaneity between people and the mammals, and published his results in a series of books and papers beginning in 1829 (e.g., de Serres, 1829, 1830, 1838). Unlike others who made similar discoveries, however, de Serres argued that the associations implied not the antiquity of the human lineage, but the recency of the extinctions, an argument that met with no success and that de Serres himself was later forced to abandon (de Serres, 1855–1857).

Paul Tournal took a markedly different approach to the same data. A pharmacist from the small town of Narbonne, just north of the eastern foothills of the Pyrenees, Tournal's interest in cave excavations was probably gained directly from de Serres, from whom he took courses at Montpellier (L. Dulieu, personal communication, 1983). Tournal, however, was the first of the two to discover human remains in an ancient context. In 1827, he announced the discovery of two caverns near Bize, just north of Narbonne; the following year, he announced that these caverns had now provided human bones in the same strata as those of extinct mammals (Tournal, 1827, 1828).

Like many others written by provincial amateurs during the 1820s and 1830s, Tournal's 1827 and 1828 publications on his work at Bize are simple descriptions, first of a series of caves, and then of an unexpectedly ancient context for human bones. After 1828, however, Tournal broadened his perspective significantly, and began not only to discuss the context and nature of his finds in more detail, but also to explore the wider theoretical meaning of those finds.

In 1829, Tournal used the faunas in the caverns he had excavated, and in those reported by the young paleontologist Jules de Christol (1829; Christol was also a student of de Serres), to place those sites in relative chronological order. He also took a broad approach to establishing the contemporaneity of those faunas with people. Not only did stratigraphic data support that contemporaneity, he argued, but some of the extinct mammal bones from Bize appeared to have been cut by human hands. In addition, following an argument developed by de Serres (1829) and that soon earned Cuvier's displeasure (see Grayson, 1983), Tournal maintained that the morphology of some of those animals suggested they had been domesticated.

Simply establishing the validity of the association, however, did not address the main issue: human remains in deposits indicative of a non-modern world. Tournal (1829a) tackled this question directly and in a strongly gradualistic way. People, he observed, had caused many large mammals to decrease in number during recent times, and there was no reason to think that this had not also happened in the deeper past. No change in the nature of the environment, other than the appearance of people themselves, was thus needed to account for the extinctions. Not much reading between the lines was required to infer that if this were the case, faunally based distinctions between modern and pre-modern worlds could not be easily drawn. Indeed, Lyell (1832) was soon to stress recent, human, caused extinctions in a very similar argument. In addition, Tournal argued, the diluvium graded so evenly into more recent deposits that it was often impossible to tell where one stopped and the other began. Lyell (1830) was to agree with this position as well.

This approach was remarkable. Tournal had used three separate lines of evidence to support the validity of the discoveries themselves. He had used the faunal associates of his human relics to array those relics in time. And, he had dissolved the distinction between pre-modern and modern worlds by a combination of observations of contemporary extinction processes and geological fact.

He restated much the same position in 1833, with one extremely important difference. In 1829, he had followed those geologists who equated the diluvium with the Deluge, and concluded that the existence of antediluvian peoples was "in accord with the book that forms the basis of the beliefs of European civilization" (1829a, p. 251). In 1833, he explicitly rejected the use of concepts even tangentially related to the Bible, terms that included, because of its association with the biblical Deluge, "antediluvian":

> It has always seemed to me that one should avoid the intervention of the authority of religious tradition in scientific discussion; but since there are those who have wanted to cross over the wall that separates the two and place the question on this ground, we must be allowed to follow in all liberty the thread of our observations and of our inductions (1833, p. 168).

Having eliminated the theologically tainted terminology, Tournal could no longer refer to antediluvian peoples. A new terminology was needed, and Tournal accordingly prepared a tripartite scheme for all of earth history. The Ancient Geological Period (Période géologique ancienne) referred to all of earth history prior to the appearance of the human species. The Modern Geological Period (Période géologique moderne) was defined by the human presence, and was in turn divided into two parts. The Prehistoric Period (Période ante-historique) began with the appearance of people on Earth and continued until the beginnings of the "most ancient traditions" (1833, p. 175); the beginnings of those traditions, some 7,000 years ago, also marked the beginning of the Historic Period (Période historique).

Over a period of four years, Tournal had thus reconceptualized the more recent aspects of earth history, allowing thereby a context for his cave discoveries. At the same time, he had defined a prehistory—the first such definition to appear—in order to avoid intermingling religion and science. It is hard to overemphasize the magnitude of Tournal's achievement, not because it presaged later conceptual developments, but because he saw, and was able to rework in contemporary terms, all of the major theoretical issues that otherwise stood in the way of his empirical discoveries at Bize.

Narbonne is about as far as you can get from Paris and still be in France; it is 160 km west of Montpellier and 240 km east of Toulouse, the only sizeable towns, and the only true centers of learning, then in the region. How could an isolated provincial pharmacist have accomplished what Tournal did? The Narbonne *Archives municipales* hold a small collection of Tournal's letters dating to between 1827 and 1833, the period of his cave research. It does not diminish Tournal's feat to note that these letters show that he did not do it alone. The letters document that during this time, he carried on an active correspondence with members of the French scientific elite, including François Arago, Jean-François d'Aubuisson, Adolphe Brongniart, and Pierre Cordier. These men were powerful and respected scientists; with the exception of Brongniart, who was elected in 1834, all were members of the *Académie des sciences* during these years.

It is not particularly interesting that Tournal wrote to important people and heard back. It is, however, of import that his correspondents informed him of problems in his interpretive framework and suggested, either directly or indirectly, ways in which that framework could be improved. In addition, they actively assisted him in finding outlets for his work.

For example, Tournal's descriptive and unremarkable 1828 paper on the Bize caverns was not only published in the *Annales des Sciences Naturelles,* but was also sent in letter form to the *Académie des Sciences,* and was read to that body on 9 February 1829 (see Tournal, 1829b). The next day, Cordier wrote Tournal with a list of questions raised by the audience. "How," he asked, "do you avoid the objection that human remains could have been introduced during modern times and buried, as in a cemetery, and that the fragments were perhaps dug up and dispersed by carnivores?" Tournal was thus immediately informed that the integrity of the association between human remains and those of extinct mammals was going to be a focal point. It is no surprise, then, that his post-1828 papers include nonstratigraphic criteria—cut marks and domestication—as part of his attempt to validate that association.

Outside suggestions were also crucial to Tournal's definition of the "Prehistoric Period." In 1831, he sent two papers to the *Académie Royale des Sciences, Inscriptions, et Belles-lettres de Toulouse,* one dealing with the Bize caverns, the other with the evidence they, along with Christol's sites, provided for antediluvian peoples, as Tournal then called them. D'Aubuisson was permanent secretary of that body, and evaluated the manuscripts for possible publication. He liked and published them both, along with his evaluation (Tournal, 1834). In that evaluation, d'Aubuisson noted that while he could not adopt Tournal's conclusions in their entirety, he agreed on the crucial point: "when extinct races of bears, rhinoceros, and hyaenas existed, man lived in our regions . . . it appears that the honor of having been the first to have shown this belongs to M. Tournal" (in Tournal, 1834, p. 63).

What aspects of Tournal's conclusions did d'Aubuisson dislike? That information he provided in a letter to Tournal on 28 April 1831. D'Aubuisson objected strongly to Tournal's use of several terms. "I condemn the use of the expression diluvium and of diluvian formation [*terrain diluvien*]" he told Tournal; "you conclude that in our region, man was an antediluvian being. But this conclusion conforms only with the system of M. Buckland." Most generally, d'Aubuisson attacked the use of theologically tinged terms in scientific analysis:

> It does not please me when the authority of religious traditions intervenes in our scientific discussions. A wall of separation must be built between two so very different things. Let us respect the holy books and let us follow, in full liberty, the thread of our observations and of our inductions.

That Tournal took these words to heart should be clear: he used the very same words, taken directly from d'Aubuisson's letter, to introduce his own critique of theologically related terms in his 1833 paper. It was d'Aubuisson who directly led Tournal to define a "prehistoric" period of human existence, since that term was meant to replace "antediluvian."

Tournal's 1833 contribution thus derived much of its direction from d'Aubuisson's suggestions. It derived something equally important, however, from another of Tournal's correspondents, François Arago. Director of the Paris Observatory and professor of astronomy at the Ecole Polytechnique, Arago was also permanent secretary of mathematical sciences of the *Académie.* He was also a major opponent of Cuvier (the other permanent secretary, of physical sciences, from 1803 until his death in 1832; see Appel, 1987 for an excellent discussion of the adversary relationship between the two men), and co-editor (with Gay-Lussac) of the respected *Annales de Chimie et de Physique.* Enemy of Cuvier and powerful scientist in his own right, Arago wrote to Tournal on 15 September 1832, four months and two days after Cuvier's death, hoping to "convince [him] to give me a general paper on the caverns of Bize for the *Annals of Chemistry.*" Obviously, Tournal complied, and his paper, containing so much that Cuvier would have found objectionable, soon appeared. It was Tournal's last contribution on the subject, as his interests had already turned in other directions (Rouville, 1876–1877).

A PROGRAMMATIC REJECTION

While Tournal was attempting to reconstruct contemporary views of human history in order to accommodate his discoveries, the physician Phillipe-Charles Schmerling was excavating a series of caves to the north, near Liège, Belgium. Although Schmerling was initially interested in the paleontological content of these sites, several provided not only the remains of extinct mammals, but human bones and artifacts as well.[1] Schmerling published his results in 1833 and 1834, and, like Tournal and de Serres, argued for contemporaneity. Indeed, Lyell visited Liège in 1833, and reported to the geologist Gideon Mantell that Schmerling had "found human bones in breccia, embedded with the extinct species, under circumstances far more difficult to get over than any I have previously heard of" (Lyell, 1881, v. 1, p. 402).

But get over them he did, as did all other members of the scientific elite until late in the 1850s. At first, isolated reports of human bones and artifacts associated with the remains of extinct mammals could be treated on a case-by-case basis (e.g., Cuvier, 1812; Buckland, 1823). As the years went on, however, the number of cases grew to the point that a more detailed and more

[1]One of Schmerling's sites, Engis, provided the badly broken skull of a child that Schmerling was unable to reconstruct. Huxley (1863) failed as well, and it was not until 1936 that the reconstruction was accomplished and the skull recognized as Neanderthal (Fraipont, 1936). Even had the skull been whole, however, the course of intellectual history would have remained unaltered, since odd morphology did not mean great time depth, and a variety of explanations, including that the skull was pathological, would have explained the discovery.

encompassing explanation for their existence became essential. The simple enumeration of asserted associations never led to a change in attitude concerning human antiquity, but it did lead directly to the development of a programmatic rejection of cave data. Since virtually all the associations reported through the mid-1840s came from caves, the development of such a programmatic rejection applied to all known cases. That rejection was provided in 1832 by Charles Lyell in Britain and by Jules Desnoyers in France.

Lyell's *Principles of Geology,* the first edition of which appeared between 1830 and 1833, viewed earth history as nondirectional and nonprogressive, changing while remaining the same. To maintain this view, he had to dismiss those aspects of the fossil record that appeared to demonstrate direction and progression, and this he did by arguing that all such evidence was produced by gaps in the fossil record and in our knowledge of it. Given this system, Lyell had two choices when it came to dealing with the issue of human antiquity. He could have treated the human species as just another kind of organism, and argued that our species was truly ancient. In the late 1820s, this position could have been supported by using the data presented by Tournal and his colleagues, but he could have made our species even far older, attributing the lack of extremely ancient human remains to the incomplete and poorly explored fossil record. Taking that position would have avoided the difficulty of dealing with the abrupt appearance of people on an Earth whose history was otherwise strictly uniform.

Lyell, however, was acutely aware that the notion of the progressive development of life through time held within it the seeds of the view that people were merely another, albeit the highest, animal, the current endpoint in life history (Wilson, 1970; Bartholomew, 1973). To Lyell, such a conclusion threatened the special status of our species by denying that we had originated as a result of "special, independent creative attention" (Bartholomew, 1973, p. 284; see also Bowler, 1976). Lyell was interested not in making our species old and ordinary, but in making it young and distinct. Thus, he argued that our lineage had only recently appeared on Earth, and as a "considerable departure from the succession of phenomena previously exhibited in the organic world," a "new and extraordinary circumstance" (1830, p. 156). This was a position that could not be reconciled with Lamarck's, or anybody else's, tranformationist accounts of life history.

In the second volume of the *Principles,* Lyell dealt with the evidence from southern France, which had suggested to some that people had coexisted with the now-extinct mammals. Did it, he asked, document such coexistence? His answer left no doubt as to his position: "we should unquestionably have arrived at this conclusion if the bones had been found in an undisturbed *stratified* deposit . . . but we must hesitate before we draw analogous inferences from evidences so equivocal as that afforded by the mud, stalagmites, and breccias of caves, where the signs of *successive* deposition are wanting" (1832, p. 225–226). The sediments of caves, he argued, are so easily and so frequently disturbed that "it is not on such evidence that we shall readily be induced to admit either the high antiquity of the human race, or the recent date of certain lost species of quadrupeds" (1832, p. 226–227). This powerful conclusion appears in every edition of the *Principles* from the first to the ninth (1853).

Desnoyers (1831–1832), then secretary of the Geological Society of France, took a slightly more eclectic approach to his dismissal of cave data. He observed that the kinds of artifacts that had been reported in association with the remains of extinct mammals included some that were very similar to those found in such clearly recent sites as dolmens. Such stylistic similarity, he concluded, must imply temporal similarity as well, and the human remains from the bone caverns hence must also be recent. Like Lyell, he made the simple and correct observation that the deposits of caves are prone to disturbance from a broad variety of agents, from people to water, and the disturbances so caused can readily produce accidental associations between recent and ancient objects. "The essential fact," Desnoyers observed, "is to show intimate connections, to prove that no subsequent cause could have operated to mix debris and animals whose existence could have been separated by a long period of time" (1831–1832, p. 251). It was just this intimate association in an undoubtedly undisturbed context that had yet to be demonstrated, and that could not be shown in sites with such "discordant stratigraphy" (1831–1832, p. 252) as caves. This argument fell on extraordinarily receptive ears. Not only was it soon translated and published in English (Desnoyers, 1834), but Lyell immediately incorporated it into the *Principles* as support for his own position: Desnoyers' argument appears in every edition of the *Principles* from the third (1834) to the ninth (1853).

The arguments made by Lyell and Desnoyers may have been arguments in the service of a particular belief, but they were nonetheless sound. It was well known at the time that cave stratigraphy is often treacherous. The merit was so great, in fact, that Tournal himself admitted it, while maintaining that the associations on which he based his position were valid. Tournal also realized the threat that Desnoyers' position posed to the use of cave data in addressing issues relating to human antiquity, amounting, he said, to an announcement "*a priori* that the problem of contemporaneity that has been discussed for so long can never be resolved" (1831–1832, p. 381; see also Tournal, 1833).

Tournal was correct in thinking that acceptance of arguments such as Desnoyers' meant that cave data would become inadmissable evidence for a deep human antiquity. That, in fact, is precisely what happened, and Tournal's own impressive accomplishments were to play no further role in establishing the reality of a deep human antiquity. He was wrong, however, in thinking that there was no other source of information that could be brought to bear on the issue.

A NEW SOURCE OF DATA

During the early 1840s, just such a new source began to be developed by Jacques Boucher de Perthes, director of customs in

Abbeville, northwestern France. Men such as Tournal, de Serres, Christol, and Schmerling excavated in caves because of the paleontological content of the sediments that filled them; their discoveries of human remains in those sediments were fortuitous by-products of their paleontological interests. Boucher de Perthes' work, however, had been stimulated by Casimir Picard, who, during the 1830s, had published on the stone and antler tools found in the valley of the Somme near Abbeville (e.g., Picard, 1834–1835, 1836–1837). Boucher de Perthes adopted Picard's interest in artifacts, but combined it with an attention to stratigraphy that was derived from the work of local geologists (e.g., Ravin, 1834–1835). Further, he followed these local colleagues with an interest not in cave deposits, but instead in those of the Somme River Valley, deposits that were deep, often well stratified, and, in places, paleontologically rich. In the right hands, this combination of interests could become powerful.

Boucher de Perthes rewrote his own history liberally (Aufrère, 1940). According to his own account, he found his first ancient artifact in 1840 at Menchecourt, then made similar finds in 1844 at the Moulin Quignon and L'Hôpital sites. A prolific author in many realms, he did not hesitate to publish his results; the first volume of his *Celtic and Antediluvian Antiquities* appeared in 1847 (see Grayson, 1983, for a discussion of the publication history of this volume). Here, Boucher de Perthes emphasized that the deposits in which he had found his artifacts had also provided the remains of extinct mammals: elephant and rhinoceros at Menchecourt, elephant at L'Hôpital and Moulin Quignon. He not only emphasized the stratigraphic integrity of the thick deposits in which he worked, but he also provided measured stratigraphic profiles for Menchecourt and L'Hôpital that indicated the precise positions from which his artifacts had come.

Like Tournal, Boucher de Perthes recognized that his discovery was full of deep implications concerning our place on Earth. Like Tournal, he attempted to formulate an earth history that accommodated that discovery. Unlike Tournal, he did not have the knowledge and connections to carry it off. He embedded his archaeological evidence in a theoretical framework that was far more similar to the outmoded speculations of eighteenth century *philosophes* than it was to any contemporary scientific approach. He argued that earth history had been punctuated by massive catastrophes, each of which had annihilated all life and after which new life forms had been created. While this in itself was a standard form of catastrophism still acceptable in France, it was most certainly not standard to argue that the human species was among the destroyed organisms, and that this had happened "as many times as there were geological revolutions" (1847, p. 496). After these destructions, people had been recreated, but because each destruction had altered the surface of the Earth, the newly recreated peoples had to be of slightly different form than the preceding versions. He was certain that if ancient human bones were someday to be found, "nuances of form" (1847, p. 245) shown by those bones would prove him correct. The earliest people, he argued, had first appeared thousands of centuries ago, a view that was mild compared to the millions of centuries he had penned into the draft of the first volume of *Celtic and Antediluvian Antiquities* (Aufrère, 1940)[2]. Other aspects of Boucher de Perthes' work appeared equally preposterous from a contemporary perspective. No wonder Darwin, on reading his work at the time, "concluded that the whole was rubbish" (Darwin, 1887, v. 3, p. 15–16).

No wonder also that Boucher de Perthes' first volume met with no success. Undaunted, he continued working, his second volume appearing in 1857. Now he was able to turn to the support his work had been given by Marcel-Jérôme Rigollot, a physician from the Somme Valley town of Amiens, who, in 1853, had replicated the Menchecourt, L'Hôpital, and Moulin Quignon data at the site of St. Acheul (Rigollot, 1854). In addition, Boucher de Perthes retooled his approach to earth and life history, presenting views that were no longer a century out of date. The second volume, however, did not contain the crucial stratigraphic information on Menchecourt and L'Hôpital. That information was found only in the first volume, where it was surrounded by extremely odd statements on earth and life history. As a result, it was difficult to separate the old Boucher de Perthes from the new one.

THE QUESTION RESOLVED

In January, 1858, limestone quarrying operations near Torquay, southwestern England, revealed the existence of a paleontologically rich cave. The geologist and educator William Pengelly visited the site, and agreed to cooperate with Hugh Falconer, a paleontologist, in its excavation. To Falconer, the value of Brixham Cave lay in the information it might provide on the sequence of Pleistocene faunal change in England (Prestwich, 1873), and he appealed to the Geological Society of London for financial support to carry out the needed work. The Society responded not only by assisting him in attaining that funding, but also by appointing a committee of respected British scientists, including Lyell, the geologist Joseph Prestwich, and the paleontologist Richard Owen, to oversee the project. This was, as Gruber (1965, p. 385) has noted in his superb discussion of these events, a cave "to be explored by *geologists* with a view toward the solution of certain *geological* problems."

Brixham Cave was thus to be excavated for the same general reasons that Tournal, de Serres, and Schmerling had excavated their caves. But the differences far outweigh the similarities, differences that resulted from the stature and expertise of the scientists involved, and from the detailed nature of the questions posed. Since those questions dealt with faunal change through time, Brixham Cave was dug with exquisite care, strata being

[2]On 20 May 1940, the Germans destroyed much of Abbeville, including the Musée Boucher de Perthes and most of its contents. Included among those contents were Boucher de Perthes' manuscripts and collections (see Agache-Lecat, 1976, 1978; for a photographic account of the destruction, Anonymous, 1947). Aufrère (1940) is indispensable not only because of the insights it contains, but also because it is based on Boucher de Perthes' now-destroyed papers.

peeled horizontally, and the position of critical objects measured in.

The first flint implement associated with the remains of extinct mammals was found here on 29 July 1858. By the time the excavations ended, in the summer of 1859, 36 flint objects had been found, of which 15 were felt to be undeniably artifactual in nature. Many of these had been found in sealed strata that also contained the remains of such extinct mammals as elephant, rhinoceros, cave bear, and hyena. The initial results of these excavations were first made public at the 1858 meeting of the British Association for the Advancement of Science (BAAS), both Falconer and Pengelly presenting papers on those results. Neither was published in the *BAAS Reports,* but, as Pengelly noted, "all the great geologists" (Gruber, 1965, p. 390) were there (for Falconer's paper, see Murchison, 1868).

Brixham Cave was raising the issue of human antiquity in a way it had not been raised before. An excavation made with unparalleled care under the direction of some of Britain's finest geologists and paleontologists had discovered undoubted artifacts associated with the remains of extinct mammals in a sealed context. Nonetheless, Brixham Cave was a cave, and as Prestwich (1866, p. 214) himself said, "cave evidence alone was not sufficient." The real significance of this work was that it led immediately to the detailed examination of what it was Boucher de Perthes and Rigollot really had.

The visits to the sites at Abbeville and Amiens began with Falconer in October 1858. Prestwich, Lyell, the geologist R. C. Godwen-Austin, and the antiquarians John Evans and John W. Flower came during the spring and summer of 1859, as did the French paleontologist Albert Gaudry. All came away convinced, and all reported their conviction in 1859 to an impressive variety of scholarly bodies: Prestwich to the Royal Society of London in May (Prestwich, 1860, 1861); Evans to the Society of Antiquaries in June (Evans, 1860); Flower to the Geological Society of London in the same month (Flower, 1860); Lyell to the BAAS in September (Lyell, 1860); Gaudry to the *Académie des Sciences* in September and October (Gaudry, 1859a, 1859b). All agreed that the artifacts found at Abbeville by Boucher de Perthes and at Amiens by Rigollot established that people had coexisted with the now-extinct mammals. The human species had become ancient.

SOME CONSEQUENCES

My charge here has been to deal with the establishment of a deep human antiquity, not with the immediate consequences of that establishment. I note, however, that at the same session of the 1859 meeting of the BAAS in which Lyell announced his conversion to the new view of human antiquity, he also announced the forthcoming publication of Darwin's *Origin of Species.* The two issues—human time depth and human evolution—have not been separate since.

It was immediately recognized that while a recent origin of our lineage was incompatible with transmutationist theory, the fact of great time depth was consistent with any of the available explanations for human origins, whether evolution by natural selection, evolution driven by some other mechanism, or creation. Accordingly, while some antitransformationists fought an unsuccessful battle against a great human antiquity (e.g., Dawson, 1863; Phillips, 1860), most did not. People could have been created 100,000 years ago as well as 6,000 years ago, and everyone recognized that that was the case.

In fact, with the Pleistocene time barrier broken, only the age of the oldest known ape-like creatures limited the possible antiquity of our species. To evolutionists, those ancient primates either represented potential ancestors (as for Lyell: see Wilson, 1970), or postdated the time of human divergence from the ancestral primate stock (as for Darwin, 1871). To creationists, people might have been called into being soon after the higher primates had been so called. Given that a series of Miocene-aged higher primates were known by the mid-1850s (e.g., Lartet, 1856), pre-Pleistocene human relics became completely possible. Indeed, stone tools from both Pliocene and Miocene deposits began to be reported as early as 1863: first reported, in fact, by Jules Desnoyers (1863). Within 20 years, so many late Tertiary, and now thoroughly discounted, sites had been reported, that Gabriel and Adrien de Mortillet (1881) created a separate "Eolithic Period" to contain them. Although debate raged over their validity (Grayson, 1986), the time depth involved was philosophically neutral, since it did not bear on the mechanisms by which the human species had originated. That mechanism was to be decided on the basis of the fact that creationist and evolutionist approaches made decidedly different predictions not about the timing of the human appearance, but about the morphology of the earliest peoples.

Julia Wedgwood found our indefinitely long past dreary, but as I noted, the new view on human antiquity cut far deeper than this, since it now appeared that we had been displaced from our central place in earth history. Parkinson's dilemma, now read as the *presence* of human remains in markedly non-modern contexts, had reemerged in more powerful form. Prestwich felt that dilemma acutely, arguing not only that people had clearly existed "anterior to the surface assuming its present form, *so far as it regards some of the minor features*" (Prestwich, 1861, p. 309, emphasis mine), but also that the "special purpose" of glaciation might have been to provide a state of equilibrium for the surface of our planet, whereby "it has been rendered fit and suitable for the habitation and pursuits of civilized man" (1863, p. 52). In a remarkable letter to the geologist James D. Forbes written in 1864, William Whewell (Todhunter, 1876, v. 2, p. 435–436) provided a more telling account of what it all really meant:

> I will not conceal from you that the course of speculation on this point has somewhat troubled me. I cannot see without some regrets the clear definite line, which used to mark the commencement of the human period of the earth's history, made obscure and doubtful. There was something in this aspect of the subject, as Cuvier left it, which was very satisfying to those who wished to reconcile the providential with the scientific history of the world, and this aspect is now no longer so

universally acknowledged. It is true that a reconciliation of the scientific with the religious view is still possible, but it is not so clear and striking as it was. But it is weakness to regret this; and no doubt another generation will find some way of looking at the matter which will satisfy religious men. I should be glad to see my way to this view, and am hoping to do so soon.

ACKNOWLEDGMENTS

I thank M. Paul Henri Viala of the Narbonne *Archives municipales* for his assistance. Portions of this paper were presented during my tenure as Tate Lecturer, Institute for the Study of Earth and Man, Southern Methodist University, in November, 1985. I thank James E. Brooks, Director, and David J. Meltzer for providing me with that opportunity. I thank Robert C. Dunnell, Barbara E. Grayson, and Bruce G. Trigger for comments on my thoughts concerning the development of Tournal's ideas, and James R. Sackett and Bruce Trigger for making me think about things I had neglected. I thank Barbara E. Grayson, Léo Laporte, James R. Sackett, and Ellis L. Yochelson for their helpful reviews of this paper.

REFERENCES CITED

Agache-Lecat, M., 1976, Le Musée Boucher de Perthes à Abbeville, *in* Livret-guide de l'excursion A10: Nord-ouest de la France (Bassin de la Seine, Bassin de la Somme et Nord): Paris, Union Internationale des Sciences Préhistoriques et Protohistoriques, p. 158–161.

——, 1978, Léon Aufrère (1889–1977): Bulletin de la Société d'Emulation Historique et Litteraire d'Abbeville, v. 24, no. 4, p. 455–468.

Anonymous, 1947, Abbeville: 20 Mai 1940–3 Septembre 1944: Abbeville, Le Syndicat d'Initiative d'Abbeville et ses Environs, 64 p.

Appel, T. A., 1987, The Cuvier-Geoffroy debate; French biology in the decades before Darwin: Oxford, Oxford University Press, 305 p.

Aufrère, L., 1940, Figures des préhistoriens; 1, Boucher de Perthes: Préhistoire, v. 7, p. 1–134.

Bartholomew, M., 1973, Lyell and evolution; An account of Lyell's response to the prospect of an evolutionary ancestry for man: British Journal for the History of Science, v. 6, p. 261–303.

Boucher de Perthes, J., 1847, Antiquités celtiques et antédiluviennes; Mémoire sur l'industrie primitive et les arts à leur origine, v. 1: Paris, Treuttel and Wurtz, 628 p.

——, 1857, Antiquités celtiques et antédiluviennes; Mémoire sur l'industrie primitive et les arts à leur origine, v. 2: Paris, Treuttel and Wurtz, 511 p.

Boué, A., 1825, Mémoire géologique sur le sud-ouest de la France, suivi d'observations comparatives sur le nord de même royaume, et en particulier sur les bords du Rhin: Annales des Sciences Naturelles, v. 14, p. 125–174.

——, 1829, Ossemens humains présumés fossiles: Revue Bibliographique des Sciences Naturelles 1829, p. 150–151.

——, 1830, Compte rendu des progrès de la géologie: Bulletin de la Société Géologique de France, v. 1, p. 71–75, 94–97, 105–124.

Bowler, P. J., 1976, Fossils and progress: New York, Science History Publications, 191 p.

Buckland, W., 1820, Vindiciae geologicae; Or the connexion of geology with religion explained: Oxford, 38 p.

——, 1823, Reliquiae diluvianae; Or, observations on the organic remains contained in caves, fissures, and diluvial gravel, and on other geological phenomena, attesting the action of an universal deluge: London, Murray, 303 p.

Buffon, G., 1778, Des époques de la nature: Histoire Naturelle, Générale et Particulière, Supplément, v. 5, p. 1–254.

Burchfield, J. D., 1975, Lord Kelvin and the age of the earth: New York, Science History Publications, 260 p.

Burkhardt, F., and Smith, S., eds., 1986, The correspondence of Charles Darwin, v. 2, 1837–1843: Cambridge, Cambridge University Press, 603 p.

Burkhardt, R. W., Jr., 1977, The spirit of system; Lamarck and evolutionary biology: Cambridge, Harvard University Press, 285 p.

Christol, J. de, 1829, Notice sur les ossemens humains fossiles des cavernes de département du Gard; présentée à l'Académie des Sciences le 29 juin 1829: Montpellier, J. Martel Ainé, 25 p.

Cuvier, G., 1812, Discours préliminaire, *in* Recherches sur les ossemens fossiles de quadrupèdes, où l'on rétablit les caractères de plusieurs espèces d'animaux que les révolutions du globe paroissent avoir détruites, v. 1: Paris, Deterville, v. 1, p. 1–116.

Darwin, C., 1871, The descent of man, and selection in relation to sex: London, Murray, v. 1, 433 p.; v. 2, 475 p.

Darwin, F., ed., 1887, The life and letters of Charles Darwin, including an autobiographical chapter (third edition): London, Murray, v. 1, 395 p., v. 2, 393 p., v. 3, 418 p.

Davies, G. L., 1969, The earth in decay: A history of British geomorphology 1578–1878: New York, American Elsevier, 390 p.

Dawson, J. W., 1863, On the antiquity of man; A review of "Lyell" and "Wilson": Edinburgh New Philosophical Journal, v. 18, p. 40–65.

de Luc., J.-A., 1979, Lettres physiques et morales sur l'histoire de la terre et de l'homme: La Haye and Paris.

de Mortillet, G., and de Mortillet, A., 1881, Musée préhistorique: Paris, Reinwald, 407 p.

de Serres, M., 1829, Géognosie des terrains tertiares, ou tableau des principaux animaux invertébrés des terrains marins tertiares, du midi de la France: Montpellier and Paris, Pomathio-Durville, 276 p.

——, 1830, Sur les ossemens humains découverts dans certains cavernes du midi de la France, mélés et confondus dans les mêmes limons où existent de nombreuses espèces de mammifères terrestre, considérées jusqu'à présent comme fossiles et comme antédiluvienne: Journal de Géologie, v. 2, p. 184–191.

——, 1838, Essai sur les cavernes à ossements et sur les causes qui les y ont accumulés (3rd ed.): Paris, Baillière, 412 p.

——, 1855–1857, Des ossements humains des cavernes et de l'époque de leurs dépôts: Académie des Sciences et Lettres de Montpellier, Mémoire de la Section des Sciences, v. 3, p. 13–90.

Desnoyers, J., 1831–1832, Considérations sur les ossemens humains des cavernes du midi de la France: Bulletin de la Société Géologique de France, v. 2, p. 126–133.

——, 1834, Proofs that the human bones and works of art found in caves in the south of France, are more recent than the antediluvian bones in these caves: Edinburgh New Philosophical Journal, v. 16, p. 302–310.

——, 1863, Note sur des indices matériels de la coexistence de l'homme avec l'Elephas meridionalis dans un terrain des environs de Chartres, plus anciens que les terrains de transport quaternaires des vallées de la Somme et de la Scinc: Comptes Rendus Hebdomadaires de l'Académie des Sciences, v. 56, p. 1073–1083.

Esper, J. F., 1774, Description des zoolithes nouvellement découverts d'animaux quadrupèdes inconnues et des cavernes qui les renferment de même que de plusieurs autres grottes remarquables qui se trouvent dans le Margraviat de Bareith au de la des Monts (translated by Isenflamm, J. F.): Nuremburg, Knorr, 121 p.

Evans, J., 1860, On the occurrence of flint implements in undisturbed beds of gravel, sand, and clay: Archaeologia, v. 38, p. 280–307.

Flower, J. W., 1860, On a flint implement recently discovered at the base of some beds of drift-gravel and brick-earth at St. Acheul, near Amiens: Quarterly Journal of the Geological Society of London, v. 16, p. 190–192.

Fraipont, C., 1936, Les hommes fossiles d'Engis: Archives de l'Institute de Paléontologie Humaine 16, 52 p.

Gaudry, A., 1859a, Os de cheval et de boeuf appartenant à des espèces perdues, trouvés dans la même couche de diluvium d'où l'on a tiré des haches en

pierre: Comptes Rendus Hebdomadaires de l'Académie des Sciences, v. 49, p. 454–454.
——, 1859b, Sur les résultats de fouilles géologiques entreprises aux environs d'Amiens: Comptes Rendus Hebdomadaires de l'Académie des Sciences, v. 49, p. 465–467.
Gay, P., 1966, The Enlightenment; An interpretation: New York, Norton, 555 p.
Gillispie, C. C., 1959, Genesis and geology; A study in the relations of scientific thought, natural theology, and social opinion in Great Britain, 1790–1850: New York, Harper and Row, 306 p.
Grayson, D. K., 1983, The establishment of human antiquity: New York, Academic Press, 262 p.
——, 1984, Nineteenth-century explanations of Pleistocene extinctions; A review and analysis, *in* Martin, R. S., and Klein, R. G., eds., Quaternary extinctions; A prehistoric revolution: Tucson, University of Arizona Press, p. 5–29.
——, 1986, Eoliths, archaeological ambiguity, and the generation of "middle-range" research, *in* Meltzer, D. J., Fowler, D. D., and Sabloff, J. A., eds., American archaeology past and future: Washington, D.C., Smithsonian Institution Press, p. 77–134.
Gruber, J. W., 1965, Brixham Cave and the antiquity of man, *in* Spiro, M. E., ed., Context and meaning in cultural anthropology: New York, Free Press, p. 373–402.
Huxley, T. H., 1863, Evidence as to man's place in nature: London, Williams and Norgate, 159 p.
Lamarck, J.-B.-P.-A., 1801, Système des animaux sans vertèbres, ou tableau général des classes, des ordres, et des genres de ces animaux; présentant leurs caractères essentiels et leur distribution, d'après la considération de leurs rapports naturels et de leur organisation, et suivant l'arrangement établi dans les galeries du Muséum d'Histoire Naturelle, parmi leurs dépouilles conservées; précédé du discours d'ouverture du cours de zoologie, donné dans le Muséum National d'Histoire Naturelle, l'an VIII de la République: Paris, 432 p.
——, 1809, Philosophie zoologique, ou exposition des considérations relatives à l'histoire naturelle des animaux; à la diversité de leur organisation et des facultés qu'ils en obtiennent; aux causes physiques qui maintiennent en eux la vie et donnent lieu aux mouvemens qu'ils exécutent; enfin, à celles qui produisent, les unes le sentiment, et les autres l'intelligence de ceux qui en sont doués: Paris, v. 1, 428 p.; v. 2, 475 p.
Lartet, E., 1856, Note sur un grande single fossile qui se rattache au groupe des singes supérieurs: Comptes Rendus Hebdomadaires de l'Académie des Sciences, v. 43, p. 219–223.
Laudan, R., 1987, From mineralogy to geology: The foundations of a science, 1650–1830: Chicago, University of Chicago Press, 278 p.
Lyell, C., 1830, Principles of geology, being an attempt to explain the former changes of the earth's surface by reference to causes now in operation, v. 1: London, Murray, 512 p.
——, 1832, Principles of geology, being an attempt to explain the former changes of the earth's surface by reference to causes now in operation, v. 2: London, Murray, 330 p.
——, 1834, Principles of geology; Being an inquiry how far the former changes of the earth's surface are referable to causes now in operation (3rd ed.): London, Murray, v. 1, 420 p.; 453 p.; v. 3, 426 p., v. 4, 393 p.
——, 1853, Principles of geology; or, the modern changes of the earth and its inhabitants considered as illustrative of geology (9th ed.): London, Murray, 835 p.
——, 1860, On the occurrence of works of human art in post-Pliocene deposits: Report of the Twenty-Ninth Meeting of the British Association for the Advancement of Science, Notices and Abstracts, p. 93–95.
Lyell, K., 1881, Life, letters, and journals of Sir Charles Lyell, Bart.: London, Murray, v. 1, 475 p.; v. 2, 489 p.
Murchison, C., ed., 1868, Paleontological memoirs and notes on the late Hugh Falconer, A. M., M. D. With a biographical sketch of the author; v. 2, Mastodon, elephant, rhinoceros, ossiferous caves, primeval man, and his cotemporaries: London, Hardwicke.
Ospovat, D., 1981, The development of Darwin's theory; Natural history, natural theology, and natural selection, 1838–1859: Cambridge, Cambridge University Press, 301 p.
Parkinson, J., 1811, Organic remains of a former world; An examination of the mineralized remains of the vegetables and animals of the antediluvian world; generally termed extraneous fossils. The third volume, containing the fossil starfish, echini, shells, insects, amphibia, mammalia, etc.: London, 479 p.
Phillips, J., 1860, Life on earth; Its origin and succession: Cambridge, Macmillan, 224 p.
Picard, C., 1834–1835, Notice sur des instrumens celtiques en corne de cerf: Mémoires de la Société Royale d'Emulation d'Abbeville 1834–1835, p. 94–116.
——, 1836–1837, Notice sur quelques instrumens celtiques: Mémoires de la Société Royale d'Emulation d'Abbeville 1834–1835, p. 94–116.
Porter, R., 1977, The making of geology; Earth science in Britain 1660–1815: Cambridge, Cambridge University Press, 288 p.
Prestwich, J., 1860, On the occurrence of flint implements, associated with the remains of extinct mammalia, in undisturbed beds of a late geological period: Proceedings of the Royal Society of London, v. 10, p. 50–59.
——, 1861, On the occurrence of flint implements, associated with the remains of animals of extinct species in beds of a late geological period, in France at Amiens and Abbeville, and in England at Hoxne: Philosophical Transactions of the Royal Society of London, v. 150, p. 277–317.
——, 1863, Theoretical considerations on the conditions under which the drift deposits containing the remains of extinct mammalia and flint implements were accumulated; and on their geological age: Proceedings of the Royal Society of London, v. 12, p. 38–52.
——, 1866, On the Quaternary flint implements of Abbeville, Amiens, Hoxne, etc., their geological position and history: Proceedings of the Royal Institution of Great Britain, v. 4, p. 213–222.
——, 1873, Report on the exploration of Brixham Cave, conducted by a committee of the Geological Society, and under the superintendence of Wm. Pengelly, Esq., F.R.S., aided by a local committee; with descriptions of the animal remains by George Busk, Esq., F.R.S., and of the flint implements by John Evans, Esq., F.R.S.: Philosophical Transactions of the Royal Society, v. 163, p. 471–572.
Ravin, F.-P., 1834–1835, Mémoire géologique sur le bassin d'Amiens, et en particulier sur les cantons littoraux de la Somme: Mémoires de la Société Royale d'Emulation d'Abbeville 1834–1835, p. 143–210.
Rigollot, M.-J., 1854, Mémoire sur des instruments en silex trouvés à St. Acheul, près d'Amiens, et considérés sous les rapports géologique et archéologique: Amiens, Duval and Herment, 40 p.
Rouville, P. de, 1876–1877, Notice biographique sur M. Paul Tournal, secrétaire de la Commission archéologique et fondateur du Musée de Narbonne: Commission archéologique et litteraire de l'Arrondissement de Narbonne 1876–1877, p. 5–52.
Rudwick, M.J.S., 1972, The meaning of fossils: New York, American Elsevier, 287 p.
Rupke, N. A., 1983, The great chain of history; William Buckland and the English school of geology (1814–1849): Oxford, Clarendon Press, 322 p.
Schmerling, P.-C., 1833, Recherches sur les ossemens fossiles découverts dans les cavernes de la province de Liége, v. 1: Liège, Collardin, 167 p.
——, 1834, Recherches sur les ossemens fossiles découverts dans les cavernes de la province de Liége, v. 2: Liège, Collardin, 195 p.
Sedgwick, A., 1834a, Anniversary address 1830: Proceedings of the Geological Society of London, v. 1, p. 187–212.
——, 1834b, Anniversary address 1831: Proceedings of the Geological Society of London, v. 1, p. 281–316.
Todhunter, I., 1876, William Whewell, D. D., Master of Trinity College, Cambridge; An account of his writings with selections from his literary and scientific correspondence: London, Macmillan, v. 1, 416 p.; v. 2, 439 p.
Tournal, P., 1827, Note sur deux cavernes à ossemens, découvertes à Bire [sic], dans les environs de Narbonne: Annales des Sciences Naturelles, v. 12, p. 78–82.

——, 1828, Note sur la caverne de Bize près Narbonne: Annales des Sciences Naturelles, v. 15, p. 348–350.
——, 1829a, Considérations théoriques sur les cavernes à ossemens de Bize, près Narbonne (Aude), et sur les ossemens humains confondus avec des restes d' animaux appartenant à des espèces perdues: Annales des Sciences Naturelles, v. 18, p. 242–258.
——, 1829b, Lettre écrite aux administrateurs du Muséum d'histoire naturelle, de Paris, par M. Tournal fils. Narbonne, le 25 octobre 1828: Bulletin des Sciences Naturelles et de Géologie, v. 16, p. 20–22.
——, 1831–1832, Note sur les caverns à ossemens de Bize: Bulletin de la Société Géologique de France, v. 2, p. 380–382.
——, 1833, Considérations générales sur le phénomène des cavernes à ossemens: Annales de Chimie et de Physique, v. 62, p. 161–181.
——, 1834, Cavernes à ossemens: Histoire de l'Académie Royale des Sciences, Inscriptions et Belles-lettres de Toulouse, v. 3, p. 52–63.
Wedgwood, J. J., 1863 (published anonymously), Sir Charles Lyell on the antiquity of man: Macmillan's Magazine, v. 7, p. 476–487.
Wilson, L. G., ed., 1970, Sir Charles Lyell's scientific journals on the species question: New Haven, Yale University Press, 572 p.

Manuscript Accepted by the Society June 6, 1989

Printed in U.S.A.

Geological Society of America
Special Paper 242
1990

Origins of Quaternary-Pleistocene-Holocene stratigraphic terminology

William R. Farrand
Geological Sciences, The University of Michigan, Ann Arbor, Michigan 48109-1063

INTRODUCTION

What is the Quaternary?—the Pleistocene?—the Holocene (or Recent)? For the purposes of this symposium they are geologic time periods during which those parts of the geologic record containing human fossils were deposited. At least that is a good first-order definition that goes back to Lyell's final definition of the term *Pleistocene.* On a finer scale, however, and in light of present knowledge, the first appearance of human species and the beginning of the Pleistocene may not be contemporary. In tracking down the sources of these names and their meanings, one finds that the question of meaning has two parts: What did the terms mean originally, when defined? And, how did they evolve to their present meanings? There are several current definitions of all of these terms, and not everyone agrees which is most acceptable.

I have used original sources in the preparation of this chapter wherever possible, but some of the older European ones were not available to me. Fortunately, there are good secondary sources to fill the gaps, and I have relied on the three editions of R. F. Flint's well-known text (1947, 1957, 1971), as well as Nilsson's (1983) more recent synthesis, and Charlesworth's very comprehensive (but not absolutely complete) treatise (1957). It is significant that, even in such respectable sources, the facts are not always correctly stated, or they are stated in a misleading way, commonly in the author's effort to condense the discussions. These errors are even worse in tertiary sources, such as introductory texts.

Flint's treatment of the origin of this terminology is by far the most accurate and most readable, especially in his 1947 edition. However, in his 1971 edition, even he telescopes several bits of history, resulting in misleading statements. As in the case of Lyell's works, which I discuss below, it is worthwhile to read all three of Flint's volumes in order to track the evolution of Quaternary-Pleistocene-Glacial concepts during more than 25 years of the mid-twentieth century. It is somewhat amusing to follow Flint's cogent arguments in 1947 and 1957 for eliminating the terms *Tertiary* and *Quaternary* from formal geological nomenclature, only to find that he titles his 1971 volume *Glacial and Quaternary Geology!* Even so, in that 1971 edition, Flint still argues (p. 384) for "dropping Tertiary System and Quaternary System from stratigraphic nomenclature." This would leave the "Pleistocene series [which] would include all post-Tertiary strata, as implied by Lyell in 1839" (Flint, 1971, p. 384). Flint misquotes, however, because Lyell never suggested that the Pleistocene encompasses the Recent and certainly not in 1839, when Lyell's Pleistocene was still part of the Tertiary.

THE QUATERNARY

The concept of the Quaternary as a segment of geologic time dates back to the mid-eighteenth century, as the fourth part of Arduino's basic framework (Table 1)—Primary, Secondary, Tertiary—in which Quaternary encompassed loose, alluvial deposits (according to Nilsson, 1983, p. 15). However, Arduino is not listed in Nilsson's bibliography, and he is not even mentioned by Flint or Charlesworth. Thus, Arduino's use of the term *Quaternary* seems never to have made an impact.

Early in the nineteenth century, as surficial deposits were given greater scrutiny, a distinction was made between those related to existing rivers and lakes, and those in situations that could not be explained by the existing regime. The terms *Alluvium* and *Diluvium* were employed in a time-stratigraphic sense, apparently first by Mantell (1822): *Alluvium* for deposits of the existing regime and *Diluvium* for deposits of the Biblical deluge. These terms have persisted almost to the present day in the Germanic literature as synonyms for postglacial and glacial periods or deposits.

The Quaternary was born again more explicitly in 1829 when Desnoyers proposed the term for the formations—marine, lacustrine, and alluvial—that followed the then-recognized Tertiary formations in the Seine (or Paris) basin. Desnoyers' (1829, p. 419, 487–491) arguments are based on field work—stratigraphic and structural relations, geomorphic relations of Quaternary marine beds with present-day alluvial systems, and fossil content. Moreover, he found that the Quaternary strata commonly followed the Tertiary transitionally, not with an abrupt contact. Ironically, Desnoyers renounces the term *Quater-*

Farrand, W. R., 1990, Origins of Quaternary-Pleistocene-Holocene stratigraphic terminology, *in* Laporte, L. F., ed., 1990, Establishment of a geologic framework for paleoanthropology: Boulder, Colorado, Geological Society of America Special Paper 242.

nary as quickly as he proposes it because, on one hand, he does not want to give exaggerated importance to strata that can be seen simply as a continuation of sedimentary trends from the already established Tertiary formations (1829, p. 193, footnote 1) and because, on the other hand, he was not convinced of the exact correlation of the various facies that he identified as *quaternaire.* He says, in fact, that he was urged by his colleagues to give a name to his new formation, but because of his doubts he preferred to assign local names (the Crags in England, the Faluns in central France, etc.) rather than to encumber science with a useless name if his correlation was not proven (Desnoyers, 1829, p. 403). This is a remarkably precocious perspective on modern stratigraphic practice.

Thus, Desnoyers introduced the term *Quaternary* in 1829 and immediately decided not to use it. However, others were not so reticent. Reboul (1833) employed it to distinguish formations with fossils of living species from those (of Tertiary age) in which almost all the fossils are of extinct organisms, as follows.

Quaternary Period	Historic epoch
	Anti [sic]-historic epoch

In contrast, Morlot (1856) defined and subdivided his Quaternary on the basis of glaciation, introducing a dichotomy that has plagued geologists to the present day.

Quaternary Period	Modern epoch
	Second glacial epoch
	Diluvian epoch
	First glacial epoch

There seems to be a difference of opinion on the spelling of this term: Quaternary or Quarternary? Should there be an "r" before the "t" or not? Nilsson points out in a footnote (1983, p. 15) that the "spelling Quarternary, not infrequently seen in English texts written by foreigners, is linguistically incorrect." On the other hand, Haug (1911) states that in 1856, Morlot introduced the term *Quartaire* to correct the inappropriate *Quaternaire!* Both terms exist, according to Webster's Dictionary (Gove, 1971), but in fact *quaternary* refers to something composed of four parts—in parallel with the word *ternary,* which refers, for example, to a three-component (triangular) diagram in petrology. *Quatern* is given by Webster as the fourth part of something. Therefore, we should be known preferably as students of the Quarternary, but I suppose that it is much too late to make that change. Of course, Germanic writers have it right; their form of the word is *Quartär* (German) or *kvartär* (Swedish).

TABLE 1. EVOLUTION OF CHRONOLOGICAL TERMS

1760	PRIMARY, SECONDARY, TERTIARY, QUATERNARY	Arduino (Italy)
1822	ALLUVIUM, DILUVIUM	Mantell (Britain)
1829	*QUATERNAIRE*	Desnoyers (France)
1833	EOCENE, MIOCENE, PLIOCENE, RECENT	Lyell (Britain)
1833	*QUATERNAIRE* (sp?)	Reboul (France)
1837	*EISZEIT* (ice Age)	Schimper (Germany)
1839	PLEISTOCENE (=Newer Pliocene)	Lyell (Britain)
1856	*QUARTAIRE* (cf., *Quartär,* Ger.)	Morlot (France)
1869	HOLOCENE	Gervais (France)
1873	PLEISTOCENE (=Post-Pliocene)	Lyell (Britain)

THE PLEISTOCENE

I strongly urge all geologists to appreciate their roots by reading some of the older literature, such as the works of Lyell. There are problems, of course, of inaccessibility of certain editions and of the great length of some of the works. As Rudwick (1969) points out in his introduction to the facsimile reprint, many of us tend to regard Lyell's *Principles of Geology* as "the first 'modern' textbook of geology," whereas that was not at all Lyell's intention. Most of us are familiar with Lyell's ideas or contributions only from the introductory chapter of beginning geology textbooks, but some of these treatments are in error, or at least distort Lyell's intentions. Lyell clearly indicated that his work (in Rudwick's words) "was designed to effect a complete reorientation of geological thinking." Unfortunately, we are so hard pressed to read even the current literature that we blanch when faced with a book some 1,200 pages long, although Lyell's style is eminently readable. We blanch even more when we realize that in order to understand all of Lyell's contributions to the nineteenth century revolution in geological thinking, we must read also parts or all of the 11 (or 12) editions of his *Principles,* several editions of his *Elements of Geology* (which is more in the style of a textbook), and at least two editions of his *Antiquity of Man.* These works span some 47 years during which Lyell's thinking evolved along with that of the entire geologic community. The titles are usually cited in much shortened versions of the original titles, which in their entirety provide great insight into the author's purpose. For example, Lyell's first edition (1830–1833) is *Principles of Geology, Being an Attempt to Explain the Former Changes of the Earth's Surface, by Reference to Causes Now in Operation.*

Lyell subdivided the Tertiary on the basis of the percentage of living Mollusca represented among the fossil assemblages of western Europe, such that (at the time of his first edition in 1830–1833) his Eocene epoch had only 3.5 percent of recent forms, the Miocene 17 percent, the Older Pliocene 35 to 50 percent, and the Newer Pliocene 90 to 95 percent. Recognition, however, should also be given to the person who counted all these mollusks. Lyell collected quite a few fossils himself (Lyell, 1833, Appendix II), but the Frenchman G. P. Deshayes preceded him in this endeavor. Lyell discovered Deshayes' unpublished work during a visit to Paris in 1829 and persuaded Deshayes to allow him to publish his tabulation of Tertiary fossils, which thus appears as a 46-page appendix in volume 3 of Lyell's first edition (1833).

The same percentages of mollusks in the Tertiary epochs are repeated without change through most of Lyell's editions, even as late as 1871, although in 1839 Lyell recognized that ongoing field work was changing these numbers every year. Ultimately he expected to find complete gradation in percentages from one epoch to the next as discoveries of new fossil localities continued. In fact, in 1839 he revised the numbers for the Pliocene, such that the Older Pliocene contained from 50 to 70 percent, and the Newer Pliocene greater than 70 percent of modern species.

The 1839 work is of most direct concern in connection with the origin of the term *Pleistocene.* It is a French version of Lyell's textbook, titled *Nouveaux Éléments de Géologie.* It includes some material never published in an English-language edition, as far as I can determine. Had I not serendipitously consulted that edition and copied passages verbatim while I was teaching in France some 25 years ago, I would not be able to document the following remarks. In the final paragraph on the last page of that book Lyell renames his Newer Pliocene (now with more than 70 percent recent Mollusca) the Pleistocene, the Older Pliocene being shortened simply to Pliocene. He says that he finds the former terms, *Older Pliocene* and *Newer Pliocene,* too long and awkward:

> . . . j'ai trouvé que les termes destinés à désigner ces subdivisions étaient d'une longueur incommode, et je me suis proposé d'employer à l'avenir le mot Pliocène pour Ancien Pliocène, et de substituer à la dénomination de Nouveau Pliocène celle plus abrégée de Pleistocène, tirée du grec *pleiston,* plus, et *kainos,* récent. (Lyell, 1839, p. 621)

However, the term *Pleistocene* appears in none of the editions of Lyell's *Principles of Geology,* nor in the subsequent (English) editions of his *Elements of Geology,* for reasons that I explain below, but only with a new definition in his *Antiquity of Man.*

The state of knowledge changed considerably over the years during which Lyell was active. Human fossils in stratigraphic context with extinct mammals, as well as the gradual acceptance of Darwinian evolution and the transmutation of the human species, demonstrated the geological antiquity of mankind (Lyell, 1863). This led to Lyell's subdivision of *Post-tertiary* formations

TABLE 2. TIME SCALE ACCORDING TO LYELL (1873)

Abridged General Table of Fossiliferous Strata

1. Recent	Post-
2. Pleistocene (Post--Pliocene)	Tertiary
3. Newer Pliocene	
4. Older Pliocene	
5. Upper Miocene	
6. Lower Miocene	Tertiary or
7. Upper Eocene	Cainozoic
8. Middle Eocene	
9. Lower Eocene	
10. Maestricht Beds	Cretaceous
11. White Chalk	
etc.	etc.

into *Recent,* comprehending "those deposits in which not only all the shells but all the fossil mammalia are of living species," and the *Post-pliocene,* which are "strata in which, the shells being recent, a portion, and often a considerable one, of the accompanying fossil quadrupeds belongs to extinct species" (Lyell, 1863, p. 5–6). For Lyell, there was no geologic evidence of man existing "as a rational being before the close of the Pliocene Period" (Lyell, 1876, p. 564).

Now we come to the turn-around in Lyell's use of *Pleistocene.* Without explanation he had dropped the term almost as soon as he invented it. In 1863 he explains why: Lyell says that the term *Pleistocene* became immediately popular after its introduction in 1839 because Edward Forbes adopted it in his work "The Geological Relations of the Existing Fauna and Flora of the British Isles" (1846)—a work that I have not been able to consult. However, Lyell says in the *Antiquity of Man* (1863, p. 6) that Forbes used *Pleistocene* "almost precisely in the sense that I shall use Post-pliocene in this volume, and not as short for Newer Pliocene." Therefore, in 1863 Lyell suggests that it is best to abstain from the use of *Pleistocene* in the future.

All is not lost, nevertheless, because Lyell reverses himself in the fourth edition of *Antiquity of Man* (1873). Lyell writes: " . . . in a note to my 'Elements of Geology,' I advised such geologists as wished to retain Pleistocene to use it as strictly synonymous with Post-Pliocene. This . . . has been found so convenient that it has been very generally adopted" (1873, p. 4). Table 2 shows the time scale presented by Lyell (1873, p. 5).

I have not been able to locate Lyell's note to his *Elements of Geology.* It does not appear in *The Student's Elements of Geology* (1871) in which he uses *Post-pliocene* throughout. Conceivably the note is in the sixth edition of *Elements of Geology* in 1865 (not seen), but if so, it is odd that he reverted to *Post-pliocene* in the 1871 edition. No later edition of that work is included by Lyell in the list of his books that appears in the preface of the first volume of the eleventh edition of *Principles of Geology* (1877,

p. x). By the way, the word *Pleistocene* does not appear in his eleventh edition of *Principles,* nor does any mention of his basis for subdividing the Tertiary, although *Pliocene* (without qualification), *Newer Pliocene,* and *Post-tertiary* do appear in a few places. In contrast, *Pleistocene* appears in the titles of about half the chapters in the fourth edition of *Antiquity of Man* (1873). This state of affairs may be explained by the fact that the eleventh edition of *Principles,* which was published posthumously, was completed in 1872, the date given in Lyell's Preface to that volume, and thus preceded the 1873 edition of *Antiquity of Man.*

In summary, Lyell created the term *Pleistocene* in 1839 to replace *Newer Pliocene* for the final subdivision of the Tertiary Period (with more than 70 percent modern mollusks), and by 1863 he subdivided post-Tertiary time into *Post-pliocene* and *Recent.* However, Forbes had promulgated the use of *Pleistocene* for what Lyell now was calling *Post-pliocene,* and not as synonymous with *Newer Pliocene.* At first, Lyell recommended abandoning *Pleistocene* because of this confusion, but in 1873 or somewhat earlier advised its use for that segment of geologic time between the Pliocene and the Recent, essentially as we use it today.

TABLE 3. THE BEGINNING OF THE PLEISTOCENE/QUATERNARY

1839	Lyell:	more than 70 percent modern mollusks
1846	Forbes:	equivalent to Ice Age
1863	Lyell:	presence of human fossils and artifacts
1911	Haug:	First appearance in western Europe of *Equus, Elephas, Bos,* and maybe *Homo*
1913	Gignoux:	base of Calabrian Stage in southern Italy with 10 percent cold marine fauna; but still called Pliocene
1948	IGC, London:	formally recommended that base of Calabrian = base of Pleistocene along with equivalent Villafranchian stage
1965	INQUA:	first appearance of *Hyalinea baltica* in the type-section at Le Castella in Calabria
1977	Selli and others	first *Cytheropteron testudo* in better section at Vrica, 18 km from Le Castella

THE BEGINNING OF THE PLEISTOCENE OR QUATERNARY

Since Lyell's turn-around in 1873, everyone seems to be in agreement that the Quaternary and Pleistocene—call them eras, or periods, or epochs, it did not seem to make much difference to the early geologists—began at the same moment in geologic time. The Quaternary follows the Tertiary Period and comprises the Pleistocene and Holocene (or Recent) epochs. If so, what are the criteria for recognizing the beginning of the Pleistocene and the Quaternary?

Most of the separations of the Pleistocene from the preceding Pliocene are based on fossils, according to the venerable geological tradition. In 1839, Lyell recognized the Pleistocene on the basis of the percentage of modern mollusks. Later (1863), when he accepted the Pleistocene as post-Tertiary, he added human fossils as a distinguishing characteristic. In fact, he did not include some glacial deposits in England, such as the Cromer Forest Bed and related tills, within the Pleistocene because there was no evidence in them of the contemporaneity of man (Lyell, 1873, p. 260).

Before considering twentieth century usages, I want to emphasize that, at the same time Desnoyers, Deshayes, Lyell, and Reboul were sorting out their fossils, the independent discovery of ice ages was taking place (Table 3). Studies in the Alps and in Scandinavia during the first third of the nineteenth century culminated in Louis Agassiz' presidential address in 1837 and his book (Agassiz, 1840), as well as that of Charpentier (1841), on continental glaciation of a geologically recent date. The glacialists of that time did not relate their discoveries to the percentage of modern mollusks in the surrounding seas, but it was not long before it was realized that these geologic phenomena were contemporary, or nearly so. Forbes (1846) is credited with the first announcement of this correlation, in dealing with the character of the modern flora and fauna of the British Isles relative to glacial history. In 1856 Morlot, working in the Rhone valley, defined the Quaternary period (his *Quartaire*) on the basis of two glacial episodes, an interglacial that he called *diluvian,* and a succeeding *modern epoch* that we would call postglacial.

In the latter part of the nineteenth century, glacial studies, such as the comprehensive work of Penck and Brückner (1909) in the Alps and of Leverett (e.g., 1899) and others in the United States, seemed to dominate Quaternary research, and little effort was expended on correlation between glacial deposits and marine fossils. The reasons are obvious as far as the Alps and much of North America are concerned, because glacial deposits in these areas do not reach the sea. In Scandinavia, there are abundant intercalations of marine and glacial sediments, but they are almost exclusively of late-glacial to postglacial age. Therefore, the mollusks are 100 percent modern.

However, fossils of terrestrial vertebrates, such as frozen mammoths in Siberia and the bones of reindeer, muskoxen, and various elephants in western Europe, were well known to Lyell and others in the mid-nineteenth century. In fact, the association of human fossils and artifacts, in addition to cave drawings of these animals, played a critical role in establishing the time depth of human existence on Earth. In his grand treatise of 1911, Haug defined the beginning of the Quaternary on the basis of the first appearance in western Europe of three major mammalian genera—true horses, true elephants, and true cattle (*Equus, Elephas,* and *Bos*). Haug even ventured the suggestion that *Homo* should probably be added to that list, but no human fossils of such antiquity were then known.

The modern era of the study of Pleistocene-Quaternary stratigraphy begins shortly thereafter. In 1913, Gignoux published his dissertation on the Calabrian marine stage of southern Italy. He distinguishes the Calabrian from the preceding Plaisanchian and Astian stages of the Pliocene Epoch on the basis of the abrupt appearance in the Mediterranean of boreal elements in the fauna. About 10 percent of the fauna were species now living in the North Atlantic or Baltic Seas. Although this was a significant climatic break, there was little change in the physical sedimentation regime of the central Mediterranean. Gignoux, therefore, retained his new Calabrian within the Pliocene.

Intensive research on marine sequences and on terrestrial faunas during the next three decades led to a formal recommendation by the International Geological Congress in London in 1948 (see Congrès Géologique International, 1954, p. 215). Because glacial studies dominated Quaternary research during the first half of this century, and in view of the evidence for deteriorating climate documented by the Calabrian fauna, it was recommended that the base of the marine Calabrian stage should be recognized as the base of the Pleistocene Epoch (and of the Quaternary Period). Further, this correlation was extended to the Villafranchian land-mammal stage, said to be the continental equivalent of the Calabrian. However, no single type section for the Calabrian was specified, and not all of the cold-water fossils appeared at the same horizon in any given section. Thus, the definition was refined, and at the 1965 INQUA Congress (in Boulder, Colorado) the section at Le Castella in Calabria was proposed as the type section in which the first appearance of *Hyalinea baltica* (a foraminifer) would mark the debut of the Pleistocene Epoch (INQUA, 1965, p. 54). In the interim it had also been recognized that the Calabrian and Villafranchian stages are not fully equivalent. Only the upper ("cool") Villafranchian is correlative with the lower Calabrian.

Ever more detailed study of the sections in Calabria has led Italian workers to the conclusion that Le Castella and nearby outcrops are not satisfactory type sections because of unconformities and problems of facies. Selli and others (1977) have proposed that the section at Vrica, only 18 km from Le Castella, would be a much better type section. At Vrica, another member of the cold fauna, *Cytheropteron testudo* (an ostracode), occurs even lower (earlier) than *Hyalinea baltica,* and its first occurrence is considered to mark the beginning of the Pleistocene.

THE AGE AND IMPORTANCE OF THE PLIO-PLEISTOCENE LIMIT

The absolute age of the beginning of the Pleistocene was not a major concern for the early workers. By Lyell's time, the biblical chronology was already long forgotten as a measure of Earth history, and he and his contemporaries spoke freely of hundreds of thousands of years, at least, for the events of the Tertiary and Quaternary periods. In recent years, however, this question has given rise to much debate, which I cannot repeat in detail here. In brief, there are currently two schools of thought, one that would place the base of the Calabrian Stage about 1.6 or 1.8 Ma, and the other at 2.5 or even 3 Ma. The scarcity of datable materials in the Italian type section(s) that yield unequivocal ages leaves this question open.

Dating of the Le Castella section is based largely on circuitous correlation with deep-sea foraminiferal stages and an equivocal paleomagnetic age that relates it to the Olduvai magnetic subchron (Haq and others, 1977; Berggren and others, 1985). The Vrica section provides additional materials for dating, but a diversity of results has not led to a firm age. Dates of 2.1 to 2.5 Ma stem from K/Ar and fission-track dates of tuffs, but the possibility of an age less than 2.0 Ma is still possible (see summary in Nilsson, 1983, p. 87). A paleomagnetic study of the Vrica section suggests that the proposed Plio-Pleistocene boundary stratotype most closely coincides with the end of the Olduvai subchron, dated about 1.6 Ma (Tauxe and others, 1983). However, correlation of the Italian sections with those of northwestern Europe via the Villafranchian faunas of France and terrestrial floras seems to favor an age nearer to 2.3 Ma and correlation with the first (Praetiglian) northern continental glaciation (Suc and Zagwijn, 1983).

So far no human fossils or artifacts have been found in a context that can be stratigraphically correlated with the Plio-Pleistocene type section. (One might argue that Olduvai Gorge is an exception, if the Olduvai paleomagnetic subchron is the marker for the beginning of the Pleistocene.) Nevertheless, the former criterion of the first appearance of humans coincident with the beginning of the Pleistocene can no longer be taken as anything but a first approximation. Pliocene hominids are firmly dated between 2.9 and 3.8 Ma, and stone tools seem to be firmly dated at about 3.1 Ma in Ethiopia (Toth and Schick, 1986). Moreover, "dramatic morphological changes" were occurring in human populations between 3 and 2 Ma (Johanson and White, 1979), thus preceding (or partly coincident with) the beginning of the Calabrian Stage.

Difficulties in establishing the absolute age of the Plio-Pleistocene limit are linked to the perceived importance of this boundary in the geologic time scale. From the time of Desnoyers and Lyell to the present, it has been noted that the passage from the Tertiary to the Quaternary, or from Pliocene to Pleistocene, was generally one of transition, not of abrupt change. As mentioned above, Desnoyers preferred not to give his new term *Quaternary* an exaggerated importance, in part because of the transitional nature of the Quaternary strata above the Tertiary. Lyell (1839) speculated that the percentage of modern mollusks would eventually, with further field work, turn out to change gradationally from one Tertiary stage to the next, including the Pleistocene. Gignoux initially did not separate his Calabrian Stage from the Pliocene because it was largely transitional above the Plaisancian and Astian beds.

Flint (1947, 1957, 1971) argues elegantly that the terms *Tertiary* and *Quaternary* are outmoded and that the Pliocene-

Pleistocene limit "does not have 'period value' and can not justifiably be considered as separating units of higher ranks than epochs" (Flint, 1947, p. 202). Although he would use *Tertiary* and *Quaternary* as informal terms, Flint (1971, p. 384) proposes that "the Pleistocene Series would include all post-Pliocene strata, as implied by Lyell in 1839." (Unfortunately the reference to Lyell is misleading, presumably because Flint telescoped some 30 to 40 years of Lyell's ideas. In 1839, as I discuss above, Lyell's Pleistocene was still in the Tertiary, and his Recent epoch encompassed all post-Tertiary time.)

A number of writers argue in the same vein; a summary is given by Harland and others (1982, sect. 2.16) who raise the rank of the Quaternary and Tertiary from period to sub-era, and include a Pleistogene Period (encompassing the Pleistocene and Holocene epochs and sometimes called *Anthropogene,* especially in the USSR) in conformity with the now-usual grouping of the Tertiary epochs into Paleogene (Paleo-, Eo-, and Oligocene) and Neogene (Mio- and Pliocene) periods (Table 4). However logical their scheme may seem, Harland and colleagues realize that "tradition may well prevail." They add:

> . . . the [time] scale is not to embody, rather to calibrate, natural events and standardisation is intended to give a convenient and stable chronostratigraphic scale that will not vary with changing opinion. (Harland and others, 1982, sect. 2.16)

TABLE 4. SUGGESTED COMPREHENSIVE TERMINOLOGY FOR THE CENOZOIC ERA*

Cenozoic Era	Quaternary Sub-Era	(Pleistogene Period)	Holocene Epoch
			Pleistocene Epoch
	Tertiary Sub-Era	Neogene Period	Pliocene Epoch
			Miocene Epoch
		Paleogene Period	Oligocene Epoch
			Eocene Epoch
			Paleocene Epoch

*W. B. Harland and others, 1982

THE END OF THE PLEISTOCENE

The title of this section is borrowed from Watson and Wright (1980) because it is terribly apt, but it is not intended to strike fear in your heart that we will no longer have the Pleistocene to kick around. The question concerns the chronostratigraphic value of the Pleistocene-Holocene boundary. Is it time parallel or time transgressive?

Before plunging too far into that question, let us review the origin of the terms *Holocene* or *Recent.* Lyell defines the Recent in the first edition of *Principles* (1833, p. 52) as the era that "has elapsed since the earth has been tenanted by man." He also calls it the "actual period" and the "human epoch" and considers it distinct from the "tertiary strata." To put this in perspective, one must realize that in 1833, Lyell—and most other people—did not yet recognize that there are human skeletal remains or artifacts in prehistoric strata. Nevertheless, geologic strata of Recent age were acknowledged: for example, the growth of deltas documented by historic records and sediments covering architectural remains, such as the Temple of Serapis at Puzzuoli, that figures as the frontispiece of *Principles of Geology* from 1830 to 1877.

Lyell continued to use *Recent* throughout his writings. However, in 1869, Gervais (see Nilsson, 1983, p. 16) coined the term *Holocene* for the Recent, consonant with the terminology of the earlier Tertiary epochs, and it was proposed to the International Geological Congress in 1885. In contrast, *Recent* was maintained until 1968 as the official term of the U.S. Geological Survey (Cohee, 1968).

Definition of the Holocene Epoch has been troublesome, too. It has been defined as having totally modern fauna and flora and present-day sedimentary systems (the *Alluvium* of Germanic authors), and most commonly as referring to postglacial time. In contrast, Flint and others eliminate it as a formal term, extending the Pleistocene to the present day, as mentioned above. However, in 1969, 100 years after the term was created, the INQUA Commission for the Study of the Holocene proposed that the Pleistocene-Holocene boundary should be placed at 10,000 yr BP, measured in radiocarbon years (5,570-yr half life) and that a suitable type section should be located in an accessible, fossiliferous marine section (Hageman, 1972). Mörner (1976) and his colleagues have now proposed such a type section in a core taken from the Botanical Gardens in Gothenburg, Sweden, where the Pleistocene-Holocene boundary coincides with the Younger Dryas–Preboreal transition, based on pollen stratigraphy, and can be related to sea-level changes and dated by radiocarbon and varves. This reasonable suggestion has not yet been formally accepted, but there appears to be consensus on the 10,000-year date.

Watson and Wright, however, represent another school of thought, one that Flint and others would join without doubt. They believe "that subdivision of the Quaternary into the Pleistocene and Holocene should reflect practical usage as based on the stratigraphic record, particularly of climatic change" (p. 160). Specifically, the "Pleistocene-Holocene boundary should . . . always be placed at the level of local stratigraphic shift reflecting actual events rather than at the 10,000-year level" (Watson and Wright, 1980, p. 161). Thus, the true diachronic nature of the transition would be demonstrated from place to place around the globe. The other side of the coin is that this would be the "end of the Pleistocene" and of the Holocene as chronostratigraphic units of the geologic time scale because their limits would no longer be time parallel, as chronostratigraphic units are defined to be by the North American Stratigraphic Code (North American Commission on Stratigraphic Nomenclature, 1983). Watson and Wright argue that all Quaternary units are demonstrably diachronic and that our terminology and explanatory models should reflect this fact (see also Watson, 1983, 1985).

CONCLUSION

The preceding discussion is a highly condensed summary of more than 150 years of geological thought. Rather than summarizing it even further, I prefer to conclude with a remark or two about the nature of the study of the Quaternary evoked by comments by French researchers on the occasion of the Ottawa 1987 INQUA Congress. The translated title of their paper is "The Quaternary, an outmoded concept, a worn-out label, or a privileged period [of time]?" (Campy and Chaline, 1987). These authors review the origin and the unique character of the Quaternary: the young age of its deposits and therefore the abundance of sedimentological, paleontological, and climatic detail; the close connection with modern-day processes and biota, including humans; and the possibilities for fine-scale absolute dating that provide a resolution of time impossible for most of Earth history. These attributes have given rise to a field of specialization—Quaternary studies—and have given birth to societies around the world dedicated to the study of the Quaternary (such as INQUA, AMQUA, CANQUA, etc.). Campy and Chaline suggest, however, that this "pluridisciplinary" approach has served its purpose and now is creating a situation that is obsolete and even dangerous. The danger is the risk of isolation of Quaternarists from their natural colleagues in such areas as stratigraphy, paleontology, sedimentology, and geochronology. They visualize Quaternary studies as a sort of "ghetto" in which the members are cut off from normal interaction with others in their primary discipline. The opportunity for renewing concepts and discussing problems within the discipline are strongly reduced. In sum, the concept of a Quaternary period has served its purpose; now it is time to replace it with the notion of the Pleistocene as an integral part of the Neogene, with the Holocene as just the beginning of an interglacial. This approach would help us view the last 2 m.y. as a normal part of Earth history.

I would take issue with some of these ideas, but they do provide food for thought. I wonder, however, whether the ghetto-like nature of Quaternary studies is more strongly developed in Europe than in North America, and whether that is what provoked Campy and Chaline's remarks. When I started my studies some 30 years ago, students of the Quaternary did tend to be a breed apart—and they were mostly glacial geologists. In recent years, however, there has been a convergence on the study of the Quaternary time period (by whatever name) by scientists trained in a wide spectrum of specialities (oceanography, limnology, paleoanthropology, paleontology, etc.). I have no fears that their work on Quaternary problems will isolate them from broader problems of Earth history.

ACKNOWLEDGMENT

This manuscript has been considerably improved as a result of suggestions by R. A. Watson, whose thoughtful editing is sincerely acknowledged.

REFERENCES CITED

Agassiz, L., 1840, Etudes sur les glaciers: Neuchâtel, private published, 346 p.

Berggren, W. A., Kent, D. V., Flynn, J. J., and van Couvering, J. A., 1985, Cenozoic geochronology: Geological Society of America Bulletin, v. 96, p. 1407–1418.

Campy, M., and Chaline, J., 1987, Le Quaternaire, un concept dépassé?: Striolae (INQUA Newsletter), v. 1987, no. 1, p. 7–12.

Charlesworth, J. K., 1957, The Quaternary era with special reference to its glaciation: London, Edward Arnold, 2 vols., 1700 p.

Charpentier, J. de, 1841, Essai sur les glaciers et sur le terrain erratique du bassin du Rhône: Lausanne, Ducloix, 363 p.

Cohee, G. V., 1968, Holocene replaces Recent in nomenclature usage of the U.S. Geological Survey: American Association of Petroleum Geologists Bulletin, v. 52, p. 852.

Congrès Géologique International, 1954, Comptes Rendus, Section 12, Fascicule 15, 215 p.

Desnoyers, J., 1829, Observations sur un ensemble de dépôts marins plus récens que les terrains du bassin de la Seine, et constituant une formation géologique distincte; précédes d'un aperçu de la non-simultanéité des bassins tertiaires: Paris, Annales des Sciences Naturelles, v. 16, p. 171–214, 402–491.

Flint, R. F., 1947, Glacial geology and the Pleistocene Epoch: New York, John Wiley, 589 p.

——, 1957, Glacial and Pleistocene geology: New York, John Wiley, 553 p.

——, 1971, Glacial and Quaternary geology: New York, John Wiley, 892 p.

Forbes, E., 1846, On the connexion between the distribution of the existing fauna and flora of the British Isles, and the geological changes which have affected their area, especially during the epoch of the northern drift: Great Britain Geological Survey Memoir, v. 1, p. 336–432.

Gignoux, M., 1913, Les formations marines pliocènes et quaternaires de l'Italie du sud et de la Sicile: Annales de l'Université de Lyon, n.s. 1, 693 p.

Gove, P. B., ed., 1971, Webster's third new international dictionary of the English language, Unabridged: Springfield, Massachusetts, G. and C. Merriam, 2662 p.

Hageman, B. P., 1972, Reports of the International Quaternary Association Subcommission on the Study of the Holocene, Bulletin 6, 6 p.

Haq, B. U., Berggren, W. A., and van Couvering, J. A., 1977, The corrected age of the Pliocene/Pleistocene boundary: Nature, v. 269, p. 483–488.

Harland, W. B., Cox, A. V., Llewellyn, P. G., Pickton, C.A.G., Smith, A. G., and Walters, R., 1982, A geologic time scale: Cambridge, Cambridge University Press, 131 p.

Haug, E., 1911, Traité de géologie, 2 vols.: Paris, Armand Colin, 2024 p.

INQUA, 1965, Report of the VII INQUA Congress: Proceedings of the VII International Quaternary Association Congress, v. 24, 54 p.

Johanson, D. C., and White, T. D., 1979, A systematic assessment of early

African hominids: Science, v. 202, p. 321–330.
Leverett, F., 1899, The Illinois glacial lobe: U.S. Geological Survey Monograph 38, 817 p.
Lyell, C., 1830–1833, Principles of geology: London, J. Murray, v. 1 (1830) 511 p., v. 2 (1832) 330 p. (1833) 398+109 p.; reprinted in 1969, New York, Johnson.
——, 1839, Nouveaux éléments de géologie: Paris, Pitois-Levrault, 648 p.
——, 1863, The geological evidences of the antiquity of man (2nd American ed.): Philadelphia, Pennsylvania, G. W. Childs, 526 p.
——, 1871, The student's elements of geology: New York, Harper and Brothers, 640 p.
——, 1873, The geological evidences of the antiquity of man (4th ed.): London, J. Murray, 572 p.
——, 1876–1877, Principles of Geology (11th ed.): New York, D. Appleton and Co., v. 1 (1877), 671 p., v. 2 (1876), 652 p.
Mantell, G. A., 1822, The fossils of the South Downs, or illustrations of the geology of Sussex: London, L. Reife, 327 p.
Mörner, N. A., ed., 1976, The Pleistocene/Holocene boundary; A proposed boundary-stratotype in Gothenburg, Sweden: Boreas, v. 5, p. 193–275.
Morlot, A. de, 1856, Notice sur le Quaternaire en Suisse: Lausanne, Société Vaudoise des Sciences Naturelles Bulletin, v. 4, p. 41–45.
Nilsson, T., 1983, The Pleistocene: Dordrecht, Holland, D. Reidel Publishing, 651 p.
North American Commission on Stratigraphic Nomenclature, 1983, North American Stratigraphic Code: American Association of Petroleum Geologists Bulletin, v. 67, p. 841–875.
Penck, A., and Brückner, E., 1909, Die Alpen im Eiszeitalter: Leipzig, Tauchnitz, 1199 p.
Reboul, H., 1833, Géologie de la période quaternaire, et introduction à l'histoire ancienne: Paris, F. G. Levrault, 222 p.
Rudwick, M.J.S., 1969, Introduction to reprinted 1st edition of C. Lyell's *Principles of geology:* New York, Johnson Reprint Corporation, p. ix–xxv.
Selli, R., and 15 others, 1977, The Vrica section (Calabria, Italy); A potential Neogene/Quaternary boundary stratotype: Giornale di Geologia, ser. 2, v. 52, p. 181–204.
Suc, J. P., and Zagwijn, W., 1983, Plio-Pleistocene correlations between the northwestern Mediterranean region and northwestern Europe according to recent biostratigraphic and palaeoclimatic data: Boreas, v. 12, p. 153–166.
Tauxe, L., Opdyke, N. D., Pasani, G., and Elmi, C., 1983, Age of the Plio-Pleistocene boundary in the Vrica section, southern Italy: Nature, v. 304, p. 125–129.
Toth, N., and Schick, K. D., 1986, The first million years; The archaeology of protohuman culture, *in* Schiffer, M. B., ed., Advances in archaeological method and theory: Orlando, Florida, Academic Press, v. 9, p. 1–96.
Watson, R. A., 1983, A critique [of] chronostratigraphy: American Journal of Science, v. 283, p. 173–177.
——, 1985, Discussion; A critique of chronostratigraphy: American Journal of Science, v. 286, p. 766–767.
Watson, R. A., and Wright, H. E., Jr., 1980, The end of the Pleistocene; A general critique of chronostratigraphic classification: Boreas, v. 9, p. 153–163.

MANUSCRIPT ACCEPTED BY THE SOCIETY JUNE 6, 1989

Printed in U.S.A.

Geological Society of America
Special Paper 242
1990

Olduvai Gorge; A case history in the interpretation of hominid paleoenvironments in East Africa

Richard L. Hay
Department of Geology, University of Illinois, Urbana, Illinois 61801

INTRODUCTION

Olduvai Gorge occupies a key position in hominid paleoenvironmental interpretations in East Africa because of its content of hominid fossils and archeologic materials and the lengthy period of its investigation. The first investigation was in 1913 by Professor Hans Reck, a geologist from the University of Berlin. This expedition collected many fossils and found a human skeleton, then believed to be contemporaneous with fossil remains of extinct animals (Reck, 1914) but now regarded as very late Pleistocene. The outbreak of World War I ended German work in Olduvai Gorge, and the next expedition was in 1931 by L.S.B. Leakey and included Reck and A. T. Hopwood, a paleontologist. Stone tools and fossils were found throughout the sequence, and L.S.B. and M. D. Leakey worked on the archeology of the gorge for various periods between 1935 and 1947, establishing a lengthy cultural sequence from Oldowan through Acheulian cultural stages (Leakey, 1951). More detailed archeological studies were continued from 1951 to 1958.

In 1959, the skull of a hominid was found in deposits with tools of the Oldowan culture, the oldest known. This was a sensational discovery at the time, and the hominid was named *Zinjanthropus boisei,* to be later renamed *Australopithecus boisei.* This find intensified the search for hominids at Olduvai Gorge, and soon, remains of another hominid type, named *Homo habilis,* were found in the same part of the sequence as "Zinjanthropus." Hominid fossils have since been found at various levels in the Olduvai sequence, and the number of identified hominid fossils from Olduvai Gorge was listed as 46 in 1974 (Leakey, 1978). The total became 62 with the finding of a partial skeleton of *Homo habilis* in 1986 (Johanson and others, 1987). From 1960 to 1973, many archeologic sites from various levels were excavated and studied by M. D. Leakey (1967, 1971, 1975), making the cultural sequence at Olduvai one of the longest and most thoroughly studied of any Paleolithic sequence. Two sites were excavated in 1961 to 1962 by M. R. Kleindienst (1964, 1973). Thus the context of both skeletal and archeologic materials has provided an exceptional opportunity for interpreting hominid paleoenvironments.

The system for designating localities in Olduvai Gorge deserves comment. Sites with significant faunal remains or artifacts are given one or more initials of the person making the first discovery, followed by K (for karonga, the Swahili term for gully), or C, for cliff. As an example, MK is MacInnes Korongo. Reference will be made to sites FLK, FLK-N, and FLK-NN; all are in the same area, with FLK-N north of FLK, and FLK-NN the northernmost.

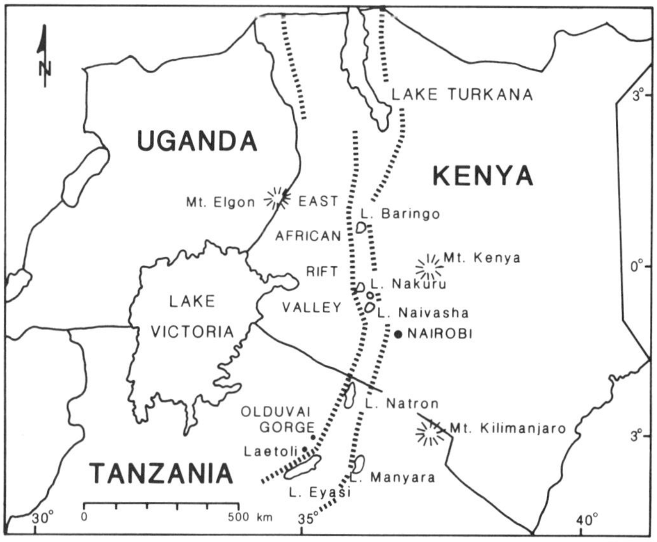

Figure 1. Map showing the part of East Africa discussed in text.

Cooke (1958) showed in detail how strongly influenced Quaternary studies in East Africa have been by the glacial-interglacial pattern of Pleistocene climates found in the Northern Hemisphere. The following brief summary is from Cooke's excellent paper. J. W. Gregory visited East Africa in 1892 to 1893 and recorded evidence for ice 5,000 to 6,000 feet below limits of existing glaciers on Mount Kenya (Fig. 1). In the eastern rift valley, known as the Gregory Rift, he documented the former existence of a series of lakes that are now dry, and he suggested that glaciation may have coincided with the maximum extension of the lakes. At this time Gregory regarded the glaciation as Pliocene, but after a second visit to Africa in 1919 he equated

Hay, R. L., 1990, Olduvai Gorge; A case history in the interpretation of hominid paleoenvironments in East Africa, *in* Laporte, L. F., ed., 1990, Establishment of a geologic framework for paleoanthropology: Boulder, Colorado, Geological Society of America Special Paper 242.

Kenya glaciation and the former extension of lakes with the Great Ice Age in Europe.

E. J. Wayland went to Uganda in 1919 convinced of a correspondence between glacial advances and "pluvial" episodes. Based to a large extent on terraces around Lake Victoria, he formed a pluvial-interpluvial interpretation of Pleistocene events in Uganda. The first East African Archeological Expedition of L.S.B. Leakey began in 1926, and early work was concentrated in the Kenya lake basins of Nakuru, Elementeita, and Naivasha, where former elevated beaches were known to occur. Solomon, in 1928, extended the geological investigations, and names were proposed for inferred pluvial and post-pluvial wet phases (Leakey and Solomon, 1929). Gregory's term Kamasian was adopted by Leakey (1931) for a pluvial stage represented by faulted deposits with handaxes at Kariandusi. These deposits were considered older than the pluvial stage represented by elevated beaches. In a later version applied to Olduvai Gorge (Leakey, 1951), the number of pluvials is expanded, and the sequence is as follows, from oldest to youngest: Kageran, Kamasian, Kanjeran, and Gamblian. In 1947 at the First Pan-African Congress on Prehistory, Leakey's terms for inferred pluvials in East Africa were accepted as stratigraphic stage names and thus appeared to give official sanction to the use of inferred climatic phases for stratigraphic and faunal correlation.

At the Pan-African Congress on Prehistory in 1955, Cooke drew attention to difficulties in correlations based on climatic interpretations (Cooke, 1957), and presented a detailed evaluation of evidence regarding climatic interpretation of the stratigraphy at different sites in East Africa (Cooke, 1958). In this he showed that most of the facts used as evidence for climatic change could be equally attributed to tectonic activity. A visit to East Africa in 1957 convinced R. F. Flint that climatic interpretations of the Quaternary features were questionable except for the relatively recent closed-basin lakes with elevated shorelines (Flint, 1959). R. Pickering, based on work at Olduvai Gorge in 1956 to 1957, also argued against correlation based on assumed stratigraphic-climatic evidence (Pickering, 1960). The climatic scheme for correlation of pre–late Pleistocene sediments was soon abandoned by all but a very few. Successful K-Ar dating of tuffs from Olduvai Gorge (Leakey and others, 1961) now revolutionized both the time scale of hominid prehistory and the basis for long-distance correlation in East Africa.

The following discussion will begin with development of the stratigraphy of Olduvai Gorge, which is essential to the history of paleoenvironmental interpretations. Paleogeographic interpretations of Beds I to IV will follow the stratigraphic discussion, to be followed by hominid paleoenvironments. The concluding section, paleoclimatic interpretation, will consider the entire stratigraphic sequence. Paleoenvironmental interpretations of the major stratigraphic units are summarized in Tables 1 to 4.

SETTING OF OLDUVAI GORGE

Olduvai Gorge is a valley at the western margin of the Eastern Rift Valley in northern Tanzania (Fig. 2). The valley is cut in the Serengeti Plain, which extends about 100 km to the northeast toward Lake Victoria. The main branch of the gorge, known as the Main Gorge, originates from Lakes Masek and Ndutu and flows eastward to empty into Olbalbal, a low-lying area in a fault graben. The Main Gorge is broad, shallow, and underlain by metamorphic rocks in the western 20 km; in the eastern 26 km it is steep sided, 45 to 90 m deep, and exposes volcanic and sedimentary rocks of Pliocene and Pleistocene age. Sedimentary deposits are also exposed in a side branch of the gorge, the Side Gorge, and along an east-west extension of the Fifth Fault to the north of the Gorge (Fig. 3). The area of Pliocene and Pleistocene sediments is bordered on the south and east by highlands comprising several large volcanoes and on the north by highlands of metamorphic rock. Several hills and inselbergs of metamorphic rock break the even surface of the plain in the vicinity of the gorge.

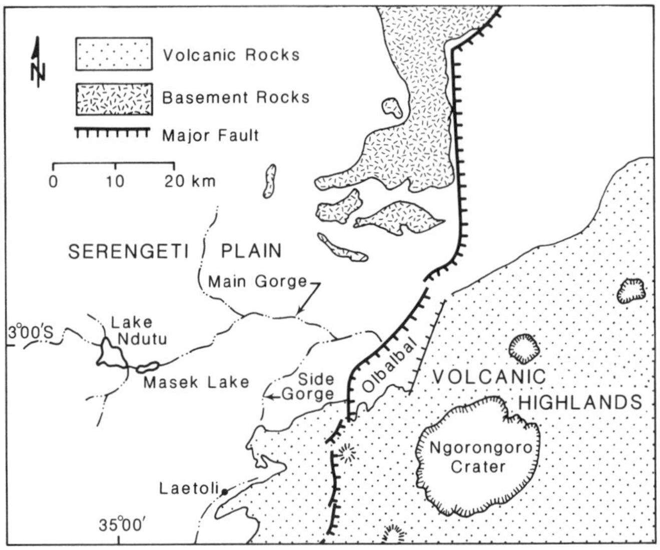

Figure 2. Regional map showing major geologic features in the vicinity of Olduvai Gorge.

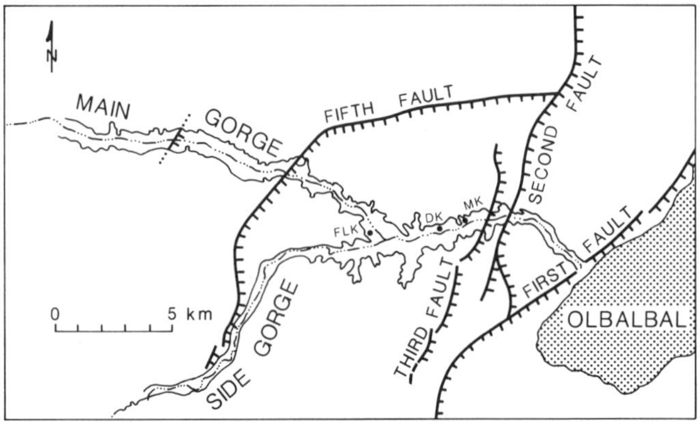

Figure 3. Outline map of Olduvai Gorge with major faults and localities referred to in text. Stippled pattern in lower right represents Holocene sediment in Olbalbal fault graben.

The Serengeti Plain is covered with grass, *Commiphora* scrub, and scattered *Acacia* (thorn trees). The bottom of the gorge contains denser vegetation, which includes much wild sisal or *Sansevieria.* The name Olduvai means "the place of the sisal" in Masai. The climate is hot and dry, and annual rainfall averages about 560 mm, nearly all of which falls from December through May. Potential evaporation vastly exceeds precipitation over the span of a year. The mean annual temperature is approximately 23°C (Hay, 1976).

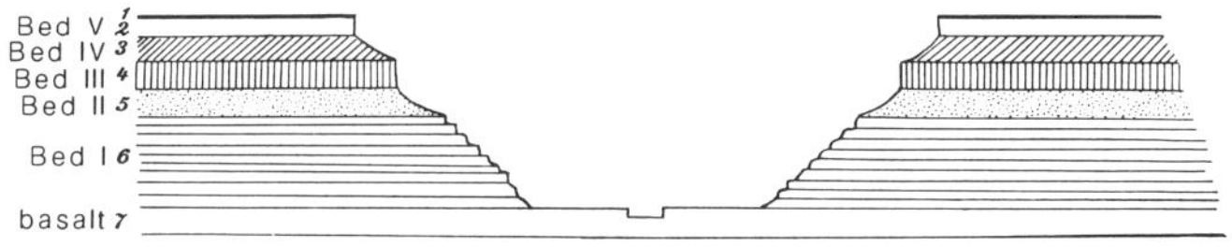

Figure 4. North–south cross section of the eastern, downstream part of Olduvai Gorge, taken from Reck (1914), Figure 2. Units are numbered from top to bottom and correspond to stratigraphic units of Reck (1951) as indicated at left. Topmost unit is a caprock of steppe limestone.

DEVELOPMENT OF STRATIGRAPHY

Development of the stratigraphic framework of Olduvai Gorge illustrates the difficulties of correlating strata in a small basin in which thicknesses and lithologies change laterally over short distances. It also illustrates the importance of tuffs in correlating and the hazards of correlating them solely on their field appearance.

Reck's first visit, for three months in 1913, established a stratigraphic subdivision for the sequence in the lower 4 km of the gorge (Reck, 1914). Basaltic lava was stated to form the igneous foundation, and the overlying sequence, 90 to 100 m thick, was described as tuffs and subdivided into units, each termed a "Schichtkomplex" or "Fossilhorizont," that were numbered from 1 to 5 (Fig. 4). Units 1 to 4 form a generally conformable sequence, and unit 5 was deposited on their deeply eroded surface.

Unit 1 was described as 30 to 40 m of cliff-forming tuffs, and unit 2 as 10 to 12 m of gray and brown fine-grained clayey tuffs. Unit 3 was described as a conglomeratic red bed, and unit 4 as clayey tuff similar to unit 2. Unit 5 was characterized as loesslike tuff. Reck noted a localized tuffaceous sandy unit above unit 4, which he was unable to assign to unit 4 or 5. He commented on, but did not study, facies changes to the west. In 1931 Reck extended his stratigraphic observations about 5 km to the west, approximately to the junction of the Main and Side gorges. In his subsequent report (Reck, 1951), he refers to the sequence as the Olduvai Beds, and his stratigraphic subdivisions are numbered as in 1914, although Roman numerals are now used, and the term Bed replaces Schichtkomplex. The units were designated "bed (1)", "bed (2)", and so on, by Leakey (1931, p. 15–17), and the earliest usage of Bed I, Bed II, and so on, with Roman numerals appears to be in Wayland (1932).

Following his visit in 1931, Reck described Bed I as a sequence of tuffs between lava flows and lacustrine marls of Bed II, and its thickness was given as 40 m near the mouth of the gorge and 6 to 8 m at the bifurcation of the gorge (Reck, 1951). Reck had unwittingly posed a problem in the definition of Bed I, as the topmost tuff near the bifurcation of the gorge was later shown to lie within the 40-m sequence of tuffs to the east.

The next stratigraphic investigation was made by R. Pickering of the Tanganyika Geological Survey, who in 1956 to 1957 mapped quarter-degree sheets that included Olduvai Gorge. His subsequent short report on Olduvai Gorge (Pickering, 1960) roughly delimited the areal extent of the Olduvai Beds and showed that they were underlain by volcanic rocks in the western part of the gorge, and he correlated the sequence in the eastern 9 km of the gorge with that exposed 4 km to the west at the Fifth Fault (Fig. 5), where lithologies differ considerably from those to the east. The western sequence he interpreted as consisting of Beds II and IV and inferred that Beds I and III pinched out to the west.

A new definition of Bed I was implied by Leakey and others (1961) in reporting the first K-Ar dates from Bed I. A marker tuff was taken as the uppermost stratum of Bed I at sites FLK and MK, which are 4 km apart. The marker tuff at FLK is the uppermost prominent tuff (Fig. 6), whereas the inferred marker tuff at MK lies near the middle of the tuff sequence considered by Reck to constitute Bed I. This use of a marker tuff to define Bed I

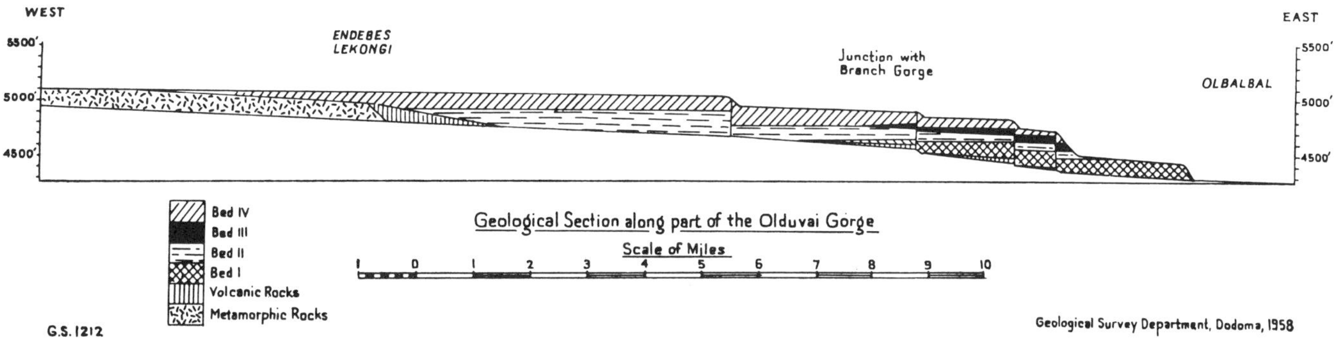

Figure 5. Geological section along the Main Gorge, taken from Pickering (1960), Figure 8. The Fifth Fault is the westernmost fault, near center of section.

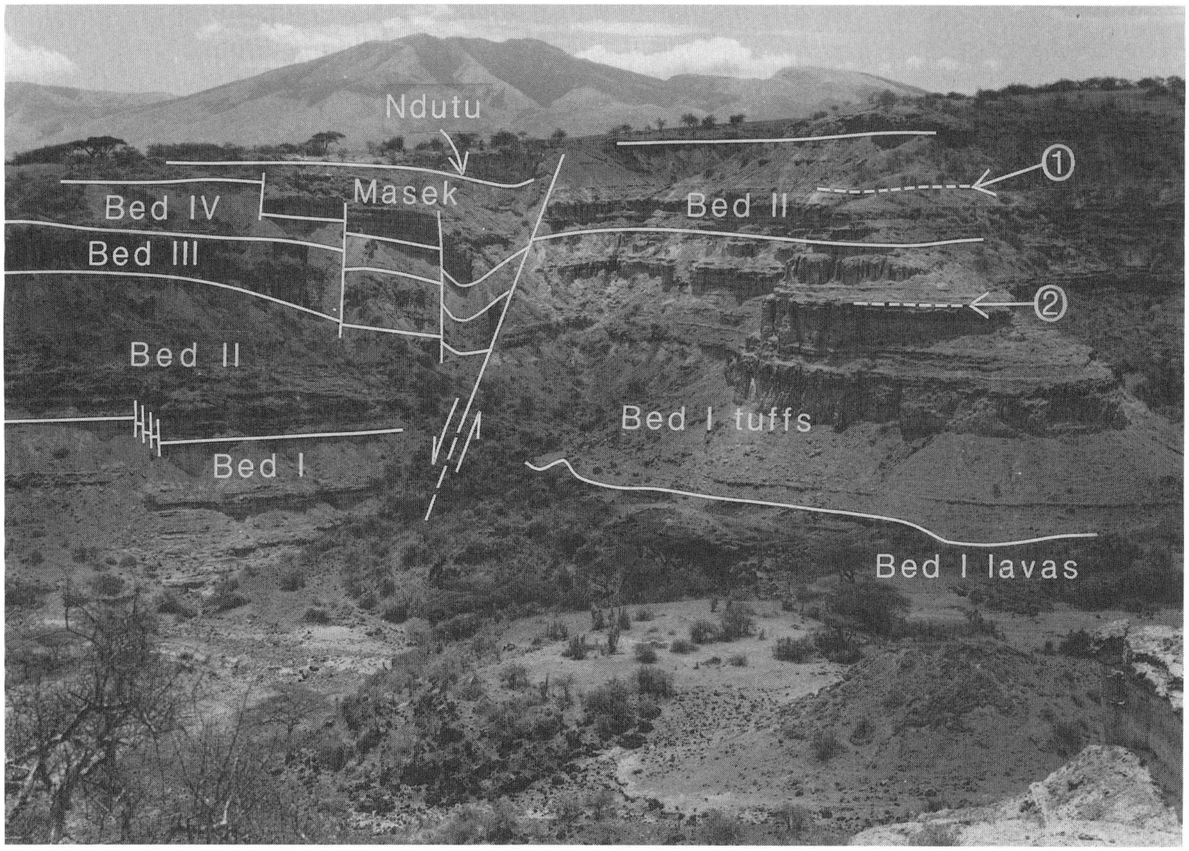

Figure 6. Photograph of the excavation at site FLK. Arrow 1 marks the discovery site of "Zinjanthropus" and arrow 2 indicates the marker tuff taken as the topmost stratum of Bed I by Reck (1951). Leakey and others (1961), and Hay (1963a). The upper 8 m of Bed I are exposed in excavation, and basalt flows are just outside photo at lower left.

changed the concept of Bed I from a deposit of tuffs to a time-equivalent unit of potentially highly variable lithology. Thus Bed I as defined by Reck would disappear where the sequence of tuffs interfingers with another lithology, whereas Bed I based on a marker tuff could consist largely of claystone.

Pickering in 1961 spent some months at Olduvai Gorge, making a geological map of the gorge and a more detailed map of site FLK. He measured a number of stratigraphic sections and wrote a report on the stratigraphy that was supplied to R. L. Hay. The report has not been published, possibly in part because of Leakey's objections (see Leakey, 1965, p. 1). His correlation of the FLK marker bed to the downstream area including MK was the same as Leakey's. Because of lateral variation and difficulties in correlating farther to the west, Pickering (1962) abandoned Reck's subdivisions and lumped Beds I and II together as the lower beds and Beds III and IV as the upper beds. He subdivided these into lithofacies designated Tuffs, Clays, Alternating Beds, Gray Beds, the Red Bed, and others.

Kleindienst made a significant stratigraphic observation in 1961 to 1962 by showing in a series of trenches that deposits of Bed III interfinger with sediments previously considered to represent Bed IV (Kleindienst, 1964, 1973).

R. L. Hay first went to the gorge in 1962 with the primary goal of determining the stratigraphy of Bed I in the eastern part of the gorge and assessing the validity of the K-Ar dates, which were highly controversial at that time because of the greatly increased time span they indicated for human evolution. Tracing of the tuffs eastward to the Second Fault (Fig. 7) showed that the marker bed of the FLK area, taken by Reck (1951) and Leakey and others (1961) as the topmost stratum of Bed I, lies within Bed I of Reck (1914). However, it was found to lie stratigraphically 12 to 15 m above the tuff taken as the marker bed by Leakey and others (1961) in the downstream area including MK and the Second Fault. L.S.B. Leakey did not readily accept this conclusion, and in a letter to Hay dated July 12, 1963, stated that the Second Fault area is where he and Reck started using the term marker bed, and there ". . . the deposit which Reck and I took as the marker bed at the top of Bed I is quite definitely present and very clearly identifiable, but equally I am certain that it is the same bed as Mary and I call the marker bed at the FLK area." Curiously, Reck (1951) does not refer to a marker bed, perhaps because of the preliminary nature of this report.

The marker-bed definition was accepted by Hay (1963a) for the top of Bed I, and the marker tuff at FLK was taken as the

Figure 7. Photograph of the south side of the Main Gorge at the Second Fault. Stratigraphy is from Hay (1976). Arrow 1 marks the top of Bed I of Reck (1914, 1951), and arrow 2 marks the top of Bed I of Leakey and others (1961). Solid line between arrows 1 and 2 follows the FLK marker bed, which marks the top of Bed I of Hay (1963a). Using this marker bed, that part of Bed I above the lavas is 38 m thick, Bed II is 20 m thick, Bed III is 13 m thick, and Bed IV is 8 m thick.

topmost stratum of Bed I (Fig. 6). The lower limit of Bed I was expanded to include the basalt flows and underlying tuffs that had not been exposed during Reck's visits. A K-Ar date of >4.0 Ma had been obtained from the lavas, but pahoehoe crusts on the topmost flow led by Hay to doubt a long hiatus between the flows and the overlying tuffs. Redating of the basalt flows gave an age of about 1.8 Ma, approximately the same as the overlying tuffs (*in* Hay, 1963a), thus supporting inclusion of the lavas within Bed I.

Two stratigraphic units characterized by wind-worked, or eolian, tuffs were recognized in the eastern part of the gorge. One in Bed II was later named the Lemuta Member, and the other, which overlies deposits of typical Bed IV lithology, was named the Masek Beds. The latter unit of eolian tuff is very likely the tuffaceous sandy deposit that Reck (1914) was unable to assign to units 4 or 5.

The major remaining problem was correlation to the area west of the Fifth Fault, where the stratigraphic section showed few points of similarity to that exposed 4 km to the east (Fig. 8). Pickering was unable to correlate between FLK and the Fifth Fault during his work in 1961, as noted above. Hay (1963a) correlated the FLK marker bed with a tuff near the base of the section, and a thick sequence of overlying claystones was assigned to Bed II. Bed III was tentatively correlated with the overlying deposits consisting largely of tuffaceous sandstone. Bed IV was correlated with a topmost claystone-rich unit. These correlations proved to be largely erroneous, and the key to correct correlations was identification of the FLK marker tuff on the basis of laboratory study of feldspar crystals in the major tuffs. The FLK marker tuff was found to lie near the top, rather than near the base of the claystone sequence to the west of the Fifth Fault (Hay, 1967). Beds III and IV were found to be a single, indivisible unit termed Beds III–IV where sediments of Bed III changed to a non-red Bed IV lithology. Beds III–IV were then correlated to the west of the Fifth Fault (Hay, 1967). Bed II in the western area was correctly delimited only after Bed I and Beds III–IV were established here.

Correlation of marker tuffs within Beds I and II was the major advance in stratigraphy from 1968 to 1974. These tuffs proved essential to working out the tectonic and paleogeographic history of Beds I and II (Hay, 1973, 1976). Reference will be made to the marker tuffs of Bed I, which are designated Tuff IA to Tuff IF, from oldest to youngest (Hay, 1971). Two discontinuous tuffs and a widespread siltstone helped to establish the sequence of archeological sites in Bed IV (Leakey, 1975).

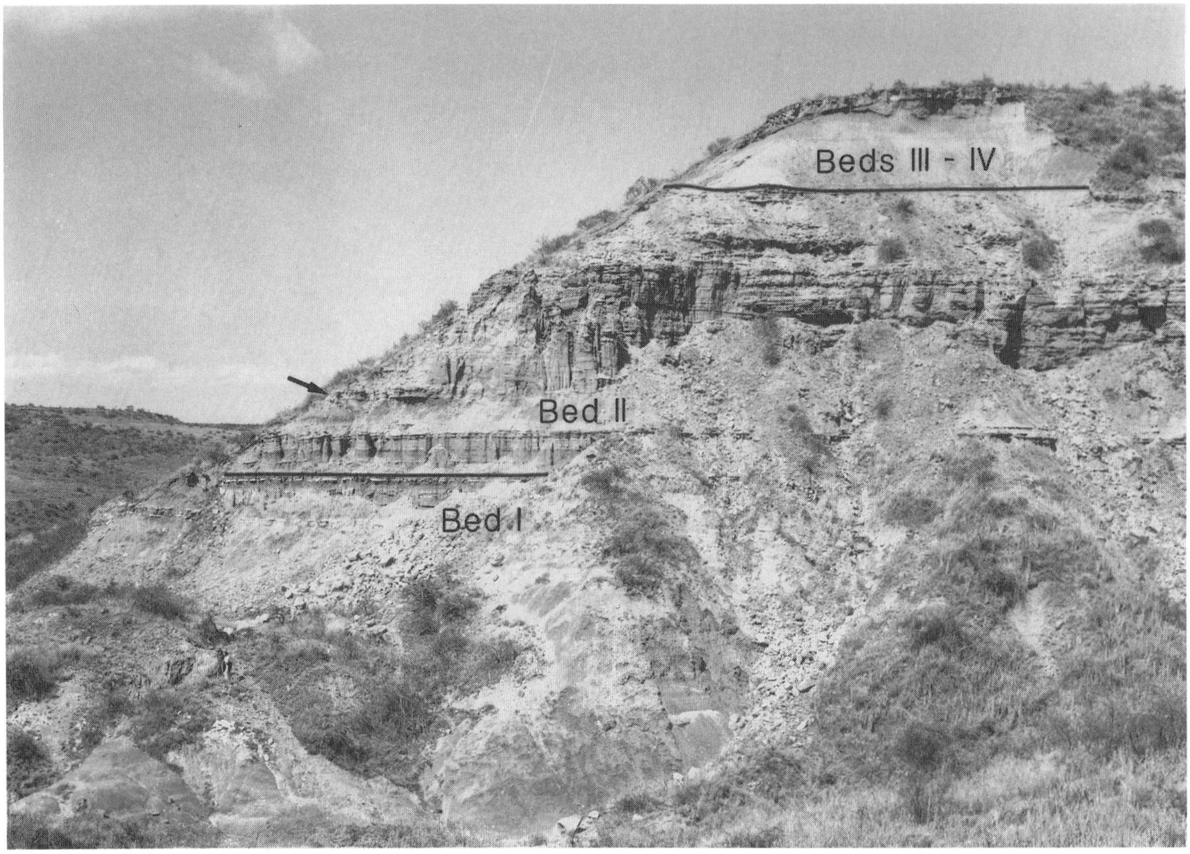

Figure 8. Photograph of sequence exposed near the Fifth Fault on the north side of the Main Gorge. Here the stratigraphic section bears little similarity to that exposed at FLK and the area to the east. Stratigraphy is from Hay (1976). Bed I is chiefly claystone and is poorly exposed; Bed II is chiefly sandstone and is 25 m thick; and Beds III–IV are chiefly claystone and are about 20 m thick. Arrow marks the top of a claystone sequence and was taken as the contact between Beds II and III by Hay (1963a). It appears to have been taken as the contact between Beds II and IV by Pickering (1960).

Discordant radiocarbon dates from Bed V (Leakey, 1969) led to its study and subdivision into the Ndutu and Naisiusiu Beds (Hay, 1971).

In summary, much labor was expended and several errors were made in developing the stratigraphic framework of Olduvai Gorge from 1914 to 1976 (Fig. 9). The major problems encountered were due to lateral variation in lithofacies and the lack of exposure of Bed I and the lower and middle parts of Bed II between FLK and the Fifth Fault. Some lateral variations are illustrated by comparing the sequence at the Second Fault (Fig. 7), with that at the Fifth Fault (Fig. 8). Bed I at the Second Fault consists of basaltic lavas overlain by tephra deposits, whereas equivalent deposits at the Fifth Fault are claystone with relatively thin tuff interbeds. Bed II at the Second Fault consists of 12 m of tephra deposits overlain by 8 m that are dominantly claystone, whereas at the Fifth Fault it comprises a lower 5 m that are dominantly claystone and an upper 20 m that are dominantly sandstone. Beds III and IV are readily recognizable at the Second Fault but are indistinguishable at the Fifth Fault.

PALEOGEOGRAPHY

Varied paleogeographic interpretations have been suggested for Beds I, II, and IV, but Bed III has been consistently viewed as alluvial (Tables 1 to 4). Reck (1914) interpreted Beds I to IV as either lacustrine or subaerial. He suggested that tuffs of Bed I are mostly subaerial, whereas a lacustrine environment was inferred for sediments of Beds II and IV on the basis of their bedding and fine grain size. A freshwater lake was postulated for Beds II and IV on their faunal content, which includes hippo, crocodile and fish. Bed III was considered subaerial on the basis of coarse conglomerates, which Reck attributed to torrential streams. He maintained a lacustrine origin for Beds II and IV after his visit to the gorge in 1931 and inferred that growth of the volcanoes caused the lake by damming eastward-flowing streams (Reck, 1951). Abundant artifacts in Bed IV led him to conclude that its lake was shallow and may have receded at times.

Wayland, director of the Geological Survey of Uganda, visited the gorge with Leakey in 1932 and disagreed with the lacustrine interpretation of the fossiliferous deposits. His views are given in a mimeographed but unpublished report entitled "The Age of the Oldoway Human Skeleton," dated September 2, 1932. In view of interest in the age of the skeleton, the report was presumably widely circulated. I received a copy of the report from Wayland in 1964. His observations are quite perceptive, and his statement is quoted in full.

> The expedition is in the habit of speaking of the "Oldoway Lake" as though the flat-lying deposits were laid down beneath a permanent sheet of water. I am not prepared to dispute that such a lake probably existed, but I contend that the Oldoway beds themselves were not deposited upon its floor, for they indicate rather the gradual filling in—to a considerable extent by windborne volcanic materials of a swampy depression, which may have been, and probably was, connected with a more permanent water area. It must be admitted that the remains of fish occur, but they are not common (as they are, for example, in the bone-beds at Kaiso, Uganda, where they are the dominant fossils), and though the relics of semi-aquatic creatures (e.g., *Deinotherium* and Crocodiles) are to be found, the fauna is essentially terrestrial, as a perusal of the lists on Fig. 3 will show. Nor were these remains carried down by rivers to be buried in deltaic deposits for these, so far as I could discover, are generally wanting. The one-time living creatures died, it would seem, pretty much where their remains are to be found this day. Some, no doubt, were killed by carnivores, and in any case it is probable that the corpses were generally dismembered after death by scavengers. Completely articulated skeletons would therefore be of a rarest occurrence; and none, I understand, has been found—except, of course that of the Oldoway man [i.e., the human skeleton found by Reck].

Leakey (1951) accepted Reck's interpretation of Beds II and IV as lacustrine and Bed III as subaerial, and he proposed that Bed I was lacustrine on the basis of crocodile, hippopotamus, and fish, and by the even horizontal stratification of its tuffs. He characterized the lakes of Beds I, II, and IV as large and shallow.

Reck (1914)	Hay (1976)		AGE (Ma)
	Major units	Subdivisions	
Unit 5	Naisiusiu Beds		0.015-0.020
Unit 5	Ndutu Beds	Upper unit	0.040-0.075
		Lower unit	0.075-0.40
Sandy tuffaceous deposit	Masek Beds	Norkilili Mbr.	0.40-0.62
		Lower unit	
Unit 4	Bed IV (Beds III - IV)		0.62-0.83
Unit 3	Bed III (Beds III - IV)		0.83-1.2
Unit 2	Bed II		1.2-1.71
Unit 1		Lemuta Mbr.	
	Bed I		1.71-1.85
Basaltic lavas		Lavas	~ 1.87

Figure 9. Comparison of the Reck's (1914) stratigraphic subdivisions of the Olduvai Beds with that of Hay (1976). Age estimates are from Leakey and Hay (1982).

His pluvial-interpluvial interpretation of stratal units very likely contributed to his oversimplifying the paleoenvironments as either lacustrine (= pluvial) or subaerial (= dry interpluvial).

Hopwood, paleontologist on Leakey's first expedition to Olduvai Gorge, viewed the faunas of Beds I to IV as not very different from the plains fauna living in the region today: "Conditions were never favorable for forest types, although they suited bush species and such animals as the hippopotamus, crocodiles, and fish" (Hopwood, 1951, p. 24). The paleogeography implied by Hopwood's faunal analysis has much in common with Wayland's view but differs considerably from the lacustrine environment proposed by Reck and Leakey.

A lacustrine origin for Beds I, II, and IV was accepted by Cooke (1958) and Flint (1959), who had visited the gorge, and by Pickering (1960, 1962), who had worked on the stratigraphy. Kleindienst's work in 1961 to 1962 may have been the first to demonstrate a non-lacustrine origin for sediments considered typical of Bed IV. She noted crossbedding in sandstones and other features, indicating that the environment was riverine or possibly deltaic (Kleindienst, 1964).

Work in 1962 led Hay (1963a) to conclude that environments for Beds I and II in the eastern part of the gorge were dominantly non-lacustrine. Bed I was interpreted as an alluvial fan of tephra deposits that interfingers westward with tuffs and claystones having features indicative of both subaerial exposure and flooding by a lake. Bed II was considered a diverse sequence of fluviatile, eolian, pyroclastic, and lacustrine sediments. A highly saline, alkaline lake was proposed for Beds I and II in the central and western part of the Olduvai basin. Evidence given for a saline alkaline lake was a lack of fossils, beds of dolomite, and authigenic K-feldspar in clays and tuffs. The modern analogy for this interpretation was Searles Lake, California, a highly saline, alkaline lake with dolomite and authigenic K-feldspar (Hay and Moiola, 1963). A fluvial origin was proposed for Bed IV on the basis of the lenticular shape of sandstones, interpreted as stream-channel deposits rather than beach sands, and on rootworkings and evidence of subaerial weathering in the associated claystones.

From 1964 to 1974, several changes were made in paleogeographic interpretations, reflecting additional stratigraphic and mineralogic data. The saline, alkaline lake was shown to be chiefly a Bed I lake (Hay, 1967), which fluctuated in level as much as 3 m and in diameter from 7 to 15 km, thus exposing a broad lake-margin zone at times of low level (Fig. 10). Molds of trona ($NaHCO_3 \cdot Na_2CO_3 \cdot 2H_2O$) and nodules of magadiite chert showed that the saline lake was rich in sodium carbonate and comparable to Lake Natron, which lies to the northeast of Olduvai Gorge (Hay, 1968). The lowermost part of Bed II was considered paleogeographically similar to that of Bed I, but the lake was shown to have been progressively reduced in size as a result of faulting and associated folding during the deposition of Bed II (Fig. 11). The area of fluvial environments and grassland increased with shrinking of the lake, thus explaining the faunal change in which remains of water-dwelling animals decrease and plains animals such as equids increase in abundance (Leakey,

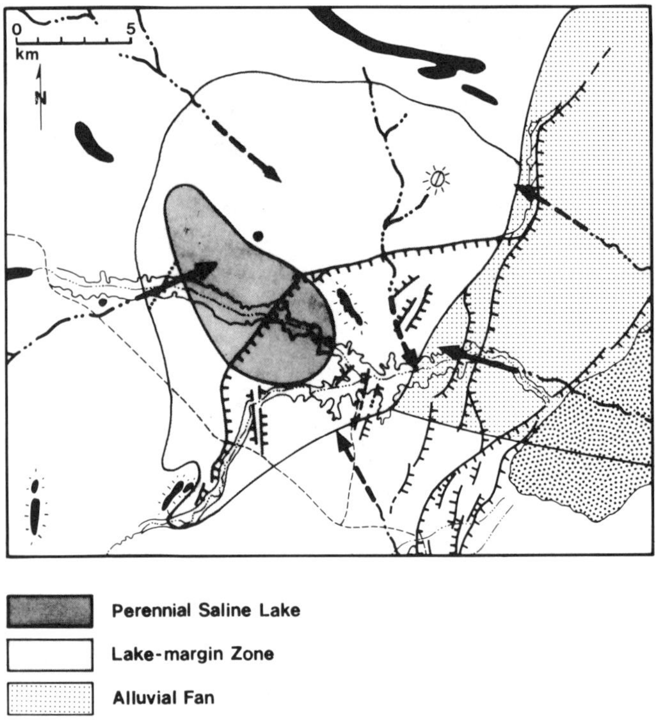

Figure 10. Regional paleogeography for the upper part of Bed I. Solid arrows indicate flow directions of streams as based on channel measurements, and dashed arrows represent flow direction inferred from clast composition and regional paleogeography. Hachured lines represent faults. Coarse stipple at lower right represents Holocene sediments. From Hay (1976), Figure 18, published by the University of California Press.

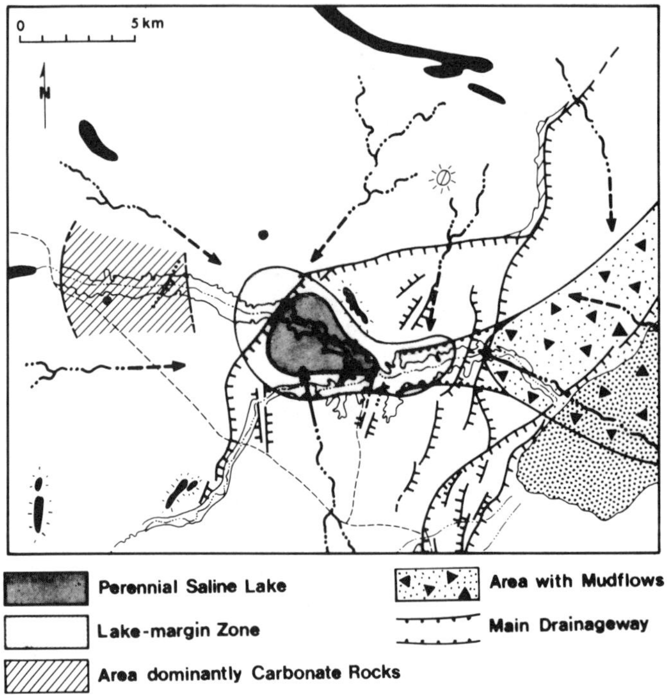

Figure 11. Regional paleogeography for a stratigraphic interval in the upper part of Bed II. For meaning of arrows and hachured lines see Figure 10. From Hay (1976), Figure 39, published by the University of California Press.

1971; Hay, 1971). After marker beds were established for Bed II, paleogeographic maps were constructed for ten stratigraphic intervals of Bed II (Hay, 1976).

Red sediments of Beds III and IV were shown to consist largely of detritus from volcanoes to the south and east of the gorge, and non-red sediments were shown to consist chiefly of detritus from Precambrian basement rocks to the north and west (Hay, 1967). Streams from the two types of terrain joined in a main trunk stream that drained eastward (Fig. 12). The trunk stream and larger streams from the west were interpreted as perennial, and streams from the volcanic highlands as intermittent (Hay, 1970, 1976). Olbalbal was first indicated as the drainage sump for streams of Beds III and IV (Hay, 1967). A playa to the north of the gorge was later proposed as a sump for streams of Bed III on the basis of a semilacustrine lithofacies (Fig. 12), and a sump to the east of the Bed III playa was suggested for Bed IV from the flow direction of Bed IV streams (Hay, 1976).

The mineral composition of Bed III and similar redbeds of Beds II and IV presented a problem in paleoenvironmental interpretation. The sandstones and conglomerates appear to be fluvial, but the claystones are extensively modified by chemical processes to zeolites, illitic clay, and iron oxide. Analcime is the principal zeolite, and phillipsite and chabazite are common (Hay, 1966, 1970). The zeolites were first thought to reflect the influence of a saline playa, inasmuch as analcimic deposits of Mesozoic age, the most nearly similar redbeds, had been attributed to saline playas or lakes (Hay, 1966). A modern example was found nearby in the basin of Lake Natron, a sodium carbonate playa lake. Here, gray and brown alluvial sediments of the Peninj Delta are altered to a zeolitic red deposit resembling Bed III in the zone intermittently flooded by saline, alkaline lake water. Based on another modern example, zeolites in claystones of Beds II to IV were later attributed to reactions in saline, alkaline soils (Hay, 1970). This second example was of fluvial sediments altered to zeolitic redbeds by saline alkaline solutions concentrated at the surface by evapotranspiration near Lake Bogoria (= Hannington) in Kenya. Beds III and IV are now reconsidered as wholly fluviatile except for small ponds represented by laminated tuffs in Bed IV.

HOMINID HABITATS

Bed I

Paleoenvironments of Bed I have received the most attention (Table 1) because Bed I contains the oldest hominid occupation sites at Olduvai and a large amount of faunal material from well-documented levels. Both *Homo habilis* and *Australopithecus boisei* are represented in Bed I, and remains of *Homo habilis* are

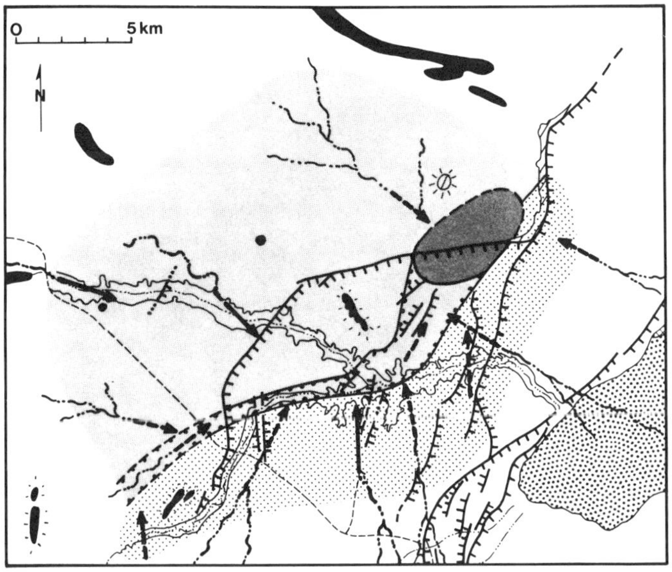

Figure 12. Regional paleogeography of Bed III with inferred drainage pattern. For meaning of arrows and hachured lines see Figure 9. The eastern fluvial facies consists largely of red volcaniclastic sediments derived chiefly from Precambrian basement rocks. From Hay (1976), Figure 45, published by the University of California Press.

TABLE 1. PALEOENVIRONMENTAL INTERPRETATIONS OF BED I

Interpretation	Reference
1. Subaerial; rainfall greater than present	Reck, 1914, 1951
2. Swampy depression	Wayland, 1932
3. Shallow lake; moist pluvial climate	Leakey, 1951
4. Dominantly plains environment	Hopwood, 1951
5. Lacustrine; rainfall greater than present	Pickering, 1960, 1962
6. Highly saline, alkaline lake with intermittently flooded lake-margin zone grading into fluvial environments; relatively dry climate	Hay, 1963a
7. Subdesertic grassland in upper most levels Bed I	Lavocat in Leakey, 1965
8. Lake rich in dissolved sodium carbonate and similar to Lake Natron, Tanzania	Hay, 1968
9. Change from wetter and more vegetated at lower levels of lake-margin zone to drier, with open grassland at higher levels	Hay, 1973
10. Rainfall higher and temperatures lower than at present	Cerling and others, 1977
11. Contraction of woodland and forest and decrease in precipitation through Bed I	Bonnefille, 1979
12. Annual rainfall 800 to 900 mm	Bonnefille and Riollet, 1980
3. Seasonal climate with long dry season	Bonnefille and others, 1982
14. Change from high-moisture and closed habitats at lowest levels to drier, open habitats at highest levels	Kappelman, 1984

the more common (Leakey, 1978). Leakey's lacustrine interpretation of Bed I (1951) implies a lake-margin hominid paleoenvironment, and hominid remains and cultural materials were inferred by Hay (1963a) to lie along the southeast margin of a saline lake in an area of streams from the volcanic highlands. Additional geological work (Hay, 1971) led to the view that the archeologic sites were situated in a broad lake-margin zone that was intermittently flooded and dried (Fig. 13).

More detailed information about Bed I hominid habitats has come from the study of paleontologic materials, principally those collected in the course of archeologic investigations by M. D. Leakey. Lakeside settings were suggested for occupation floors at DK and level 3 of FLK NN. At DK were abundant crocodile teeth and fossil rhizomes resembling those of papyrus (Leakey, 1967). This lowest archeologic site of Bed I contained a stone circle about 4 m in diameter. FLK NN yielded bones of amphibians and waterfowl and small concentrations of fish bones that may have originated by the drying of ponds cut off from the main waters of the lake (Leakey, 1971). FLK NN lies nearly 3 km west of DK, and a comparable lakeside setting is explained by the fact that an alluvial fan of tephra deposits had displaced the shoreline westward following the occupation of DK but prior to occupation of FLK NN (Fig. 13). By contrast, a "steppe or subdesertic" grassland environment was proposed for the uppermost excavated levels in Bed I at FLK N on the basis of the rodents *Gerbillus* and *Heterocephalus* (Lavocat, in Leakey, 1965). These levels contained two hominid occupation floors.

Later systematic studies of bovid, rodent, and bird remains showed a change in Bed I habitats, from wetter and more vegetated at the lowest levels to drier and with open grassland at the upper levels (Hay, 1973). The commonest bovids low in the sequence are the reduncines, a tribe containing forms that live close to water today, whereas remains of alcelaphines, characteristic of open grassland, overwhelmingly predominate at FLK N in the levels with *Gerbillus* and *Heterocephalus.* The bovid fauna of the intermediate levels at FLK was considered intermediate in character. The rodent fauna suggested a more closed environment, possibly with thickets at the lowest levels, changing to more open habitats in the highest levels. Aquatic birds were found to be more common at and below the "Zinjanthropus" horizon of FLK, and land birds (perching birds) most common near the top of Bed I at FLK N. Urocyclid slugs were found at and below the "Zinjanthropus" horizon and were interpreted to indicate damp evergreen forest where rainfall exceeded 35 inches per year or where damp conditions were maintained by regular mists. A lithofacies analysis of Bed I supported the faunal interpretation by

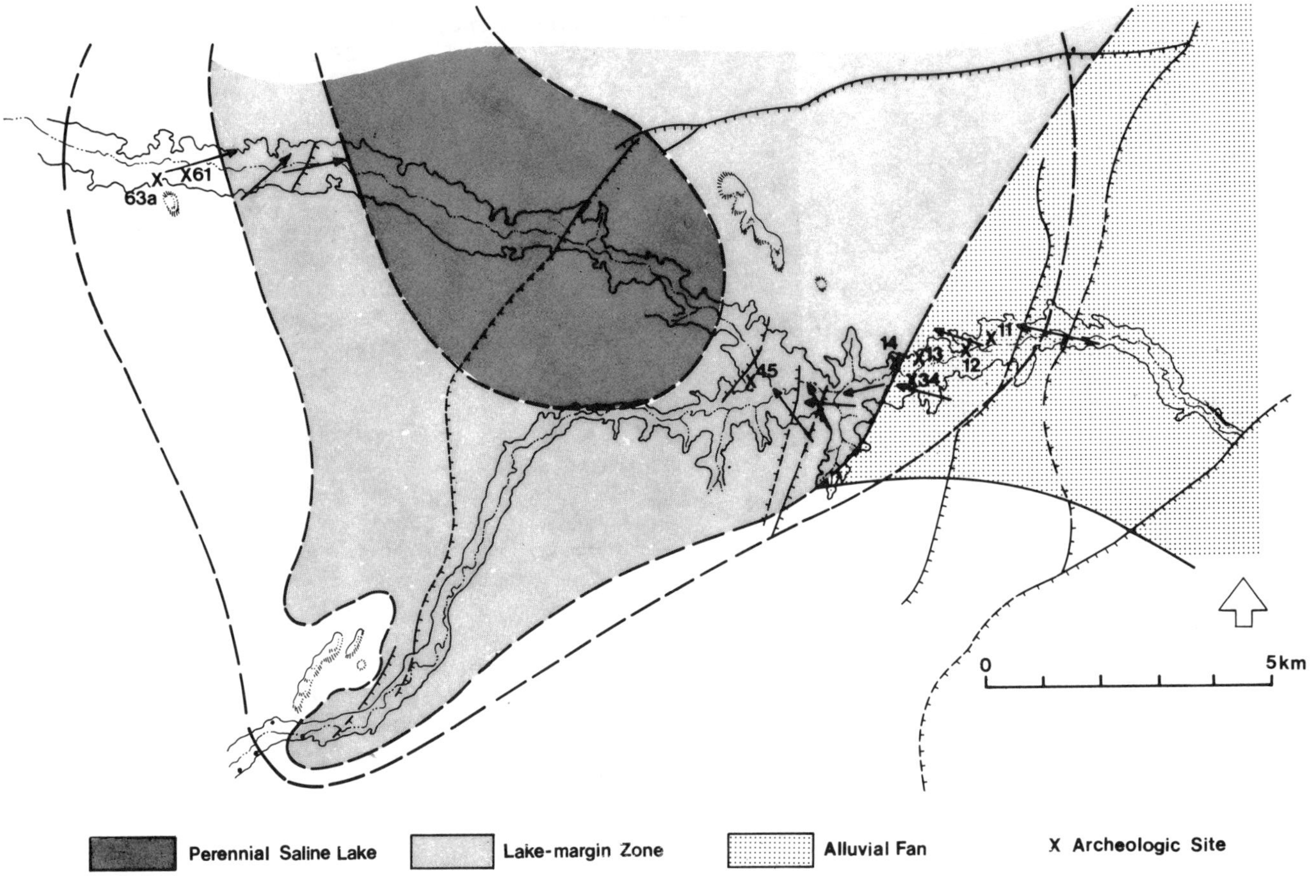

Figure 13. Paleogeography of Bed I for interval between Tuff ID and the top of Bed I. Numbered sites represent localities of artifacts, occupation sites, and hominid remains. Numbers 11 to 14 and 34 are in lake-margin deposits outlined by outer dashed line prior to episode of eruptions resulting in alluvial fan that covered the eastern 1 to 2 km of the preexisting lake-margin zone. To the west, outer dashed line shows the extent of lake-margin deposits at the level of an artifact concentration at locality 63a, which lies slightly below the levels of nos. 11 to 14 and 34. Locality no. 13 is DK, and 45 represents FLK, FLK N, and FLK NN. For meaning of arrows and hachured lines see Figure 10. From Hay (1976), Figure 17, published by the University of California Press.

showing that the lake was reduced in size while the uppermost fossiliferous deposits of Bed I were accumulating.

Pollen studies by R. Bonnefille added a new dimension to environmental interpretation of Bed I. The frequency of pollen from montane forest and woodland between the highlands and the lake records a contraction of woodland and forest and a decrease in precipitation through Bed I, with the driest, least forested conditions recorded in the uppermost levels of Bed I (Bonnefille, 1979, 1984). She suggested that temperatures of uppermost Bed I were probably higher than now. The montane flora is Sudanian-Zambezian vegetation characterized by a long dry season (Bonnefille and others, 1982) and annual rainfall between 800 and 900 mm (Bonnefille and Riollet, 1980), compared with 500 mm in the Olduvai area at present.

Kappelman (1984) interpreted the Bed I habitat from a statistical analysis of newer published taxonomic data on the bovids and murids. This was supplemented by the results of pollen studies. Kappelman's detailed analysis further documents a change from high-moisture and closed habitats for the lowest levels to drier, open habitats at the highest levels.

Data from contemporary hunters and gatherers have been used to suggest that most of the Bed I archeologic sites were occupied in the dry season; during the rainy season the Olduvai hominids either moved to nearby areas or abandoned the vicinity of the gorge as part of their seasonal migration (Speth and Davis, 1976). This interpretation is based on a comparison of faunal remains of the Olduvai sites with dry-season kills by Bushmen in the Kalahari Desert. Bushman kills divide into two distinct clusters in their proportions of Chelonia, Carnivora, and Bovidae. Tortoises are largely responsible for the clear-cut separation of the clusters. This is because land tortoises in the Kalahari hibernate in the dry season and emerge in the rainy season, when they are collected in considerable numbers. With the same three taxa, the Olduvai sites separate into two clusters almost identical to those of the Bushmen. Three Bed I sites are inferred to be rainy-season occupations, and 13 are considered to be dry-season occupations.

Potts (1984) challenged the interpretation of Bed I hominid habitats as based on artifact concentrations by postulating that the "occupation floors" of M. D. Leakey (1971) were not campsites but rather were caches of stones and stone tools left by hominids at various places in their foraging area. This interpretation leaves open questions about the size and organization of early hominid social groups and the distance that social groups lived from the sites.

Bed II

Bed II contains the largest number of known archeological sites at Olduvai Gorge and has yielded remains of three hominid types: *Homo habilis* and *Australopithecus boisei* in the lower part of Bed II, and *Homo erectus* and *Australopithecus boisei* in the upper part (Leakey, 1978).

The first habitat interpretation was of site BK, near the top of Bed II, which is rich in artifacts and faunal remains, and which had been excavated at intervals from 1952 to 1957. The site was interpreted by L.S.B. Leakey (1957) as representing a camping place of early man adjacent to a river channel or swamp with remains of many *Pelorovis,* a large bovid. He suggested that the animals had been driven into the swamp by early humans and then slaughtered. Later excavations showed that artifacts are associated with a stream channel and support the earlier scenario (Leakey, 1971).

On geologic evidence, Bed II archeologic sites were first assigned to lake-margin terrain along the southeast side of the Bed II lake (Hay, 1963a). After collaboration by M. D. Leakey and Hay, many sites in the middle and upper parts of Bed II were reported to occur in or adjacent to stream-channel deposits at variable distances from the lake, including the lower slopes of alluvial fans from the volcanic highlands (Hay, 1971).

With further stratigraphic subdivision and finer scale paleogeography, 38 of 63 Bed II sites were classed as lake margin, 18 as inland, and 7 as indeterminate (Hay, 1976). A 1-km distance from the inferred shoreline was established for classification as lake margin or inland. The excavated archeologic sites in the lower part of Bed II are classed as Oldowan (Leakey, 1971, 1975) and lie in the lake-margin zone. Faunal and floral data for the lowermost part of Bed II indicate a return to a moister, more vegetated environment than that of the upper levels of Bed I (Bonnefille, 1979; Kappelman, 1984). The fauna changes abruptly above the Lemuta Member, reflecting an increase in grassland and a decrease in both the size of the lake and area of lake-margin terrain. Middle and upper Bed II yielded coeval but dissimilar industrial complexes, termed Developed Oldowan B and Acheulian (Leakey, 1971). Nine of the ten sites classed as Acheulian are inland and one is indeterminate, whereas seven of the Developed Oldowan B sites are lake margin and two others are indeterminate (Hay, 1976). Thus the Paleolithic industry correlates closely with the paleogeography. M. D. Leakey (1971) interpreted the two complexes as "representing two distinct cultural traditions, perhaps by different groups of hominids" (1971, p. 272). Both *Homo erectus* and *Australopithecus boisei* lived in the area during this time span. An alternative interpretation, discussed but not published by G. L. Isaac, is that the two industrial complexes represent different "tool kits" used by the same type of hominid in different environments.

TABLE 2. PALEOENVIRONMENTAL INTERPRETATIONS OF BED II

Interpretation	Reference
1. Freshwater lake; rainfall greater than present	Reck, 1914
2. Swampy depression	Wayland, 1932
3. Shallow lake; moist pluvial climate	Leakey, 1951
4. Dominantly plains environment	Hopwood, 1951
5. Lacustrine; wetter than present	Pickering, 1960
6. Highly saline, alkaline lake; fluvial environments; relatively dry climate, with one episode drier than average	Hay, 1963a
7. Saline, alkaline soils during dry episode	Hay, 1963b
8. Increase in area of grassland above Lemuta Member	Hay, 1971
9. Four probable episodes of drier than average climate	Hay, 1976
10. Much woodland in lower Bed II	Bonnefille, 1979
11. Closed environments and moist climate low in Bed II	Kappelman, 1984
12. Decrease in rainfall and increase in temperature and proportion of C_4 vegetation through Bed II, with one warm, dry episode in lower Bed II	Cerling and Hay, 1986

Beds III–IV

Relatively few artifacts have been found in Bed III, whereas Bed IV contains numerous concentrations of artifacts, which have been classified as Acheulian and Developed Oldowan C (Leakey, 1975). No hominid fossils have been found in Bed III, but Bed IV has yielded remains attributed to *Homo erectus* (Leakey, 1975).

From Leakey's discussion (1951), it seems clear that he viewed the climate of Bed III as too dry for hominid occupation comparable to that of Bed IV. He noted that artifacts in Bed IV are concentrated in sandy layers, one of which he considered a beach deposit. Kleindienst (1964) showed that Bed III interfingered with non-red sediments of Bed IV lithology containing many artifacts and faunal remains, including hippo, crocodile, and fish. The same relation exists in Bed IV; non-red sediments with numerous fossils and archeologic sites interfinger southward with red volcanic alluvial deposits containing only isolated fossils and two known artifact concentrations, one of which (no. 98) may be of Bed III age (Fig. 14). Thus the sediment type, not the stratigraphic level or climate, controlled the content of artifacts and fossils in Beds III and IV. The scarcity of fossils and artifacts in deposits of Bed III lithology was later attributed to a deficiency

TABLE 3. PALEOENVIRONMENTAL INTERPRETATIONS OF BED III

Interpretation	Reference
1. Alluvial environment	Reck, 1914
2. Change from a wet to a drier and more seasonal climate	Wayland, 1932
3. Dry climate	Reck, 1951
4. Dry interpluvial climate	Leakey, 1951
5. Dominantly plains environment	Hopwood, 1951
6. Climate probably wetter than present	Cooke, 1958; Flint 1959; Pickering, 1960
7. Alluvial-playa complex and dry climate	Hay, 1963a, 1966
8. Semiarid alluvial plain with saline, alkaline soils	Hay, 1970
9. Streams braided and flow only intermittently	Hay, 1976
10. Climate drier and/or warmer than in Bed II, higher percentage of C_4 biomass than in Bed II	Cerling and Hay, 1986

TABLE 4. PALEOENVIRONMENTAL INTERPRETATIONS OF BED IV

Interpretation	Reference
1. Freshwater lacustrine; rainfall greater than present	Reck, 1914, 1951
2. Swampy depression	Wayland, 1932
3. Shallow lake	Reck, 1951
4. Lake with sand beach; moist pluvial climate	Leakey, 1951
5. Lacustrine	Pickering, 1960, 1962
6. Stream channels and flood plains; climate relatively dry	Hay, 1963a
7. Saline flood plain–playa complex	Hay, 1966, 1967
8. Semiarid alluvial plain with saline, alkaline soils; major streams contain fresh water	Hay, 1970
9 Major streams meander; small ponds in flood plain	Hay, 1976
10. Climate and percentage of C_4 biomass about the same as in bed III	Cerling and Hay, 1986

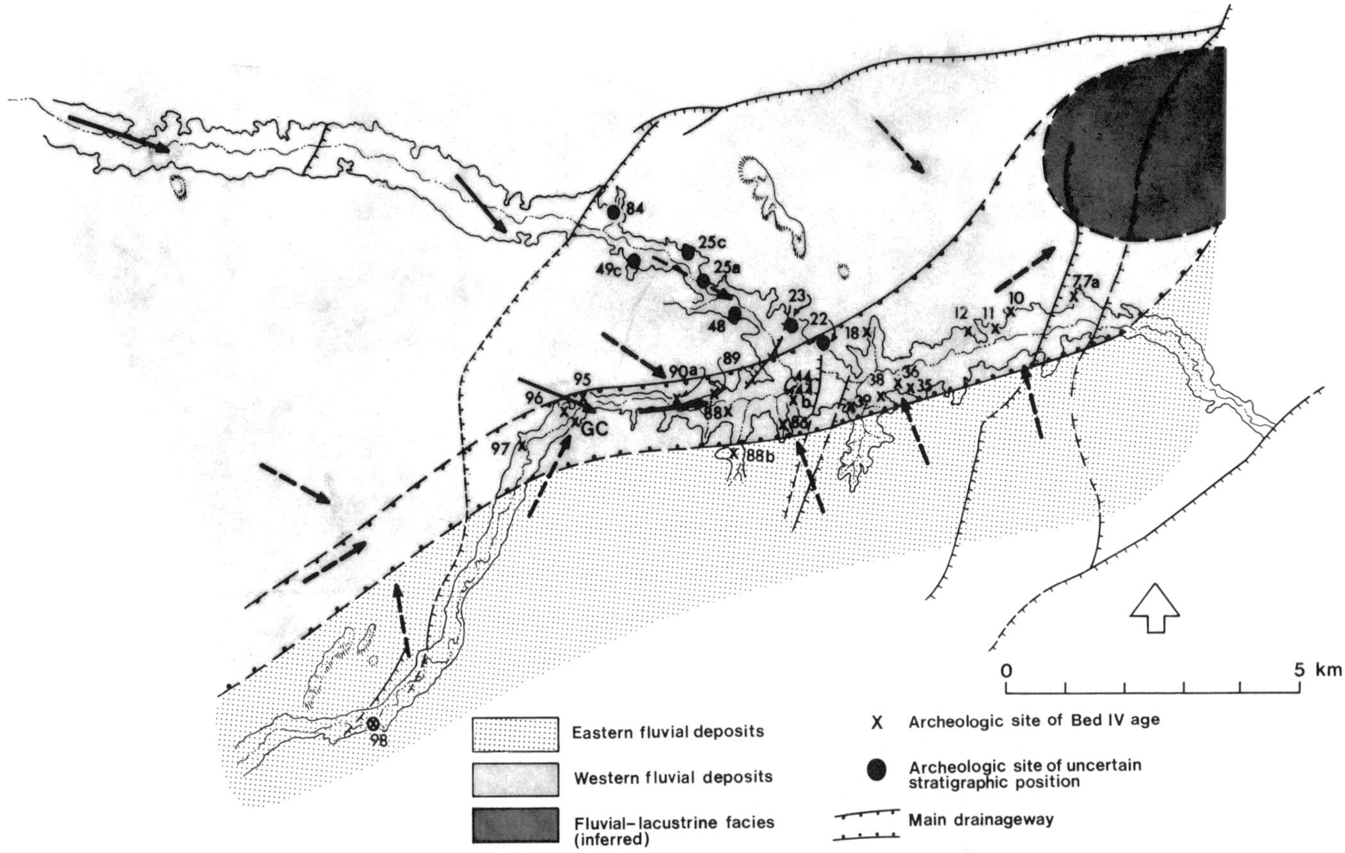

Figure 14. Paleogeography and lithologic facies of Bed IV showing localities of archeologic sites and artifact concentrations. Eastern fluvial deposits are red volcaniclastic sediments, and western fluvial deposits are non-red sediments from Precambrian terrain to the west and northwest. Sites of uncertain stratigraphic position are from Beds III–IV where the two units are lithologically indistinguishable. For meaning of arrows and hachured lines see Figure 10. From Hay (1976), Figure 44, published by the University of California Press.

of vegetation, fresh water, or both, on the Bed III alluvial surface, in contrast to the major streams depositing sediments of Bed IV type, which at least locally contained fresh water throughout the year (Hay, 1970; Kleindienst, 1973). With few exceptions, concentrations of artifacts and associated occupation debris of Beds III–IV are in sandstones and conglomerates now interpreted as stream-channel deposits (Hay, 1976). Noting the close connection between Bed IV archeologic sites and watercourses, M. D. Leakey (1975, p. 492) suggested that ". . . man may well have chosen to live in the river beds themselves during the dry season as water can often be obtained there by digging."

PALEOCLIMATE

Ideas of paleoclimate evolved from that of a climate wetter than the present for Beds I to IV to pluvials and interpluvials. The pluvial-interpluvial interpretation was criticized, partly on the questionable climatic significance of the red color in Bed III. This was followed by mineralogical and sedimentological work suggesting that the climate was relatively dry for Beds I to IV, with two episodes drier than average. An additional dry episode in Bed I was proposed on the basis of paleontological data. An overall drift to aridity from Bed I to the present has since been proposed, and inferred episodes of climatic change at 1.82, 1.67, 1.3, and 0.6 Ma have been tentatively correlated with climatic changes in the Northern Hemisphere.

Reck (1914) proposed a climate wetter than that of the present for Beds I to IV because of abundant remains of elephants, hippo, crocodile, and fish. A dry "steppe" climate was inferred for Bed V on the basis of the relative abundance of antelope remains. He suggested that the Ice Age of the Northern Hemisphere was a wet warm pluvial period in tropical Africa. Before going to the gorge, Leakey (1931) interpreted Reck's results in terms of two pluvial cycles broken by a reddened surface indicating arid conditions. Earlier recognition of elevated shorelines at Lake Manyara in Tanzania by Leakey and Solomon (Leakey, 1931) gave plausibility to this extension of the pluvial-interpluvial scheme from Kenya to Tanzania (then Tanganyika). Leakey (1951) assigned Beds I and II to the Kamasian pluvial period and Bed IV to the Kanjeran pluvial period; cutting of the pre–Bed V gorge was attributed to the Gamblian pluvial and Bed V to a subsequent dry period.

Climatic interpretations of the red color of Bed III have had a central role in both establishing and discrediting the pluvial-interpluvial interpretation of the Olduvai Beds. The earliest interpretation, by Leakey (1931), is based on the brief description of Reck (1914). Leakey states as follows: "I think that the intense reddening of bed (3), and the presence of lime concretions (kunkar?), must be taken as indications that the deposition of bed (3) was followed by a period of intense desiccation and some erosion" (1931, p. 15). Redbeds, at the time of Leakey's interpretation, were generally regarded as indicates of aridity because of the association of some redbeds with evaporites and eolian sediments. Reck (1951) also proposed that the deposits had been reddened in place, and he compared them to red-earth soils of relatively dry climates such as on the high plains of central Africa. Cooke (1958), Flint (1959), and Pickering (1960) questioned the significance of the red color as an indication of a dry climate. They were influenced by recognition of red soils in humid climates and interpretation of many redbeds as redeposited red soils (e.g., Krynine, 1950). Red soils are not presently forming on that part of the volcanic highlands draining into the Olduvai basin; hence the Bed III climate was considered to be more humid than that of the present. Hay (1963a) concluded that the red color is a result of weathering in place, and the nodular calcite is evidence of a dry climate. Illite and montmorillonite were identified as the clay minerals, indicating only slight leaching and favoring a dry climate. Subsequent work showed that Bed III is rich in zeolites as a result of the chemical reaction of volcanic detritus in saline, alkaline soils (Hay, 1970). Soils of this type were considered to require a dry climate. Thus, Bed III is currently interpreted as an arid type of redbed, as proposed by Leakey in 1931. Some redbeds elsewhere have also been reinterpreted as arid redbeds by Walker (1967), but redbeds of Bed III type differ from those described by Walker in the rapid rate of silicate reactions and reddening in saline, alkaline soils.

Although Bed III now appears to be an arid type of redbed, work on Beds I, II, and IV does not support the pluvial-interpluvial scheme. The saline lake of Beds I and II required a dry climate in which evaporation equaled or exceeded inflow and rainfall to the lake. The abrupt change from Bed II to Bed III was attributed to faulting (Hay, 1967), as suggested by Cooke (1958), rather than to a drop in lake level. The vertical transition from Bed III to Bed IV was considered a result of tectonic activity that shifted the direction of stream flow so that Bed IV sediment from Precambrian terrain was deposited over volcaniclastic sediments of Bed III in the area exposed by the eastern part of the gorge (Hay, 1963a). Drier-than-average episodes were suggested for eolian tuffs of Beds II, IV, and V. Zeolites in the eolian tuffs were attributed to weathering in saline alkaline soils, which was additional evidence of a dry climate (Hay, 1963b). Zeolitic weathering of claystones was later used as evidence for an overall dry climate for Beds III, IV, and the upper part of Bed II (Hay, 1970).

An overall increase in aridity over the past 2 m.y. was later suggested for Olduvai on the basis of faunal change and an increasing content of eolian sediment (Hay, 1976). The wettest climate was represented by the lowest fossiliferous horizons of Bed I, and eolian sediments and a paucity of mammalian remains in the Masek Beds were used as evidence that the climate was drier than that of Beds I to IV, although remains from one site showed that fresh or brackish water was available throughout the year. Catfish remains in stream-channel deposits of the lower unit of the Ndutu Beds (0.075 to 0.4 Ma) are the last evidence of a perennial supply of water in the gorge.

Isotopic measurements were tested as a paleoclimatic indicator. A preliminary oxygen-isotopic study of calcite from pedogenic carbonate limestone and ground-water nodular limestone showed an irregular increase in $\delta^{18}O$ values from Bed I to the present (Cerling and others, 1977). This was interpreted to indicate that the older carbonates were precipitated from meteoric waters at least 2 to 4‰ lower in $\delta^{18}O$ content than found in modern waters of the region. This difference was taken to imply higher humidity and rainfall and lower temperatures in the past.

After this promising start, Cerling and Hay (1986) made a systematic study of the carbon- and oxygen-isotopic composition of pedogenic and ground-water carbonates. The most significant results were on pedogenic materials, in which both $\delta^{18}O$ and $\delta^{13}C$ values increased from Bed I to the present, with "spikes" of high values in the Lemuta Member of Bed II and in the Masek Beds, and a rapid change in $\delta^{18}O$ values, accompanied by a "spike" in $\delta^{13}C$ in the transition from Bed II to Bed III (Fig. 15). Variations in oxygen-isotopic values are primarily attributed to changes in the isotopic composition of meteoric water and to a lesser extent on temperature because temperature changes would need to be unrealistically large in order to substantially affect the oxygen-isotope values. Variation in carbon isotopes is attributed to changes in the isotopic composition of soil CO_2 caused by differences in the fraction of biomass using the Hatch-Slack (C_4) photosynthetic pathway. C_4 plants yield soil CO_2 with much higher $\delta^{13}C$ values than do C_3 plants, and C_4 plants are more prevalent in drier climates than C_3 plants (Cerling, 1984). Thus, both $\delta^{18}O$ and $\delta^{13}C$ values of pedogenic calcite are indirect measures of the climate.

Cerling and Hay concluded that the record at Olduvai Gorge implies major climatic changes at about 1.67, 1.3, and 0.6 Ma, with an overall drift toward aridity. These episodes of climatic change lasted from about 50,000 to 150,000 years, and the climate after these periods of change was warmer and drier than that before. Using the pollen results, the Bed I isotope data were interpreted to suggest a mean annual temperature of about 15°C, compared to 23°C at present. The major climatic changes at 1.67, 1.3, and 0.6 Ma were tentatively correlated with those inferred from the lacustrine record in Searles Lake, California (Smith, 1984). It was also suggested that the isotopic change at 0.6 Ma may correspond to changes in global isotopic values and inferred climatic change at 0.59 Ma as based on deep-sea records.

Still another attempt has been made to correlate climatic change recorded at Olduvai Gorge with climatic change in the Northern Hemisphere. A dry episode near the end of Bed I had been documented by faunal and floral studies, as noted above, and Kappelman (1986) interpreted the paleoclimate of Bed I and lower Bed II as representing a climatic cycle from wet to dry to wet over a span of about 200,000 years. He correlated wet conditions in Bed I at about 1.82 Ma with the period of Northern Hemisphere warming, inferred from oxygen isotopes of deep-sea core V28-239. He further suggested that this moist episode in Bed I may have been a response to deglaciation, similar to high lake levels in East Africa associated with deglaciation at about 10,000 to 8,000 yrs.

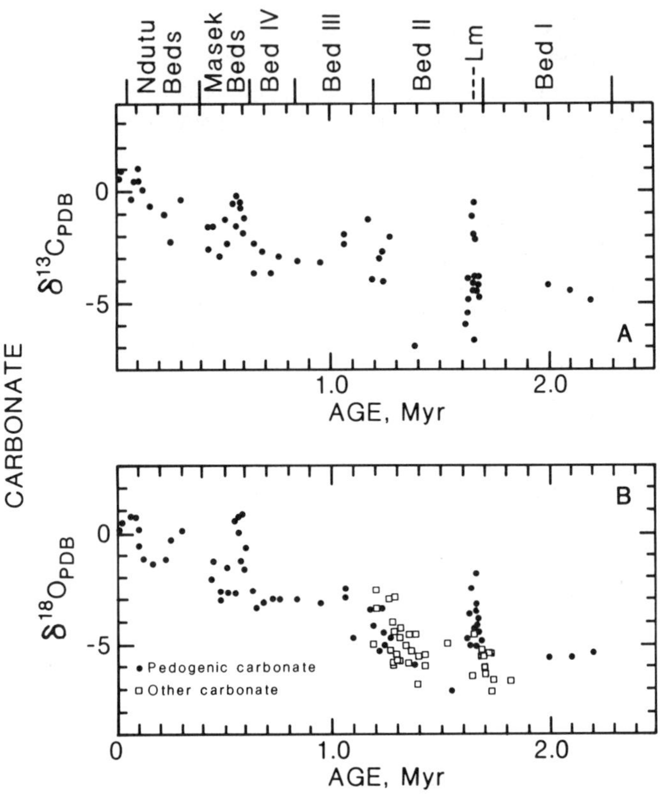

Figure 15. (A) The relationship between time and $\delta^{13}C$ for pedogenic carbonates at Olduvai Gorge. (B) The relationship between time and $\delta^{18}O$ of pedogenic and ground-water carbonates at Olduvai Gorge. Lm represents the Lemuta Member of Bed II. Modified from Cerling and Hay (1986), Figure 4, published by Quaternary Research.

ACKNOWLEDGMENTS

I am indebted to R.I.M. Campbell for the photographs used for Figures 6 to 8. Figures 10 to 14 are reproduced with permission of the University of California Press, and Figure 15 is reproduced with permission of Quaternary Research. M. D. Leakey and J. D. Clark aided in bibliographic research. The manuscript benefited considerably from reviews by H.B.S. Cooke and two anonymous reviewers.

REFERENCES CITED

Bonnefille, R., 1979, Méthode palynologique et reconstitutions paléoclimatiques au Cénozoique dans le Rift Est Africain: Bulletin de la Société Géologie France, v. 7, p. 331–342.

——, 1984, Palynological research at Olduvai Gorge: National Geographic Society Research Reports, v. 17, p. 227–243.

Bonnefille, R., and Riollet, G., 1980, Palynologie, végétation et climates de Bed I et Bed II à Olduvai, Tanzania: Proceedings, 8th Pan-African Congress of Prehistory and Quaternary Studies, Nairobi, 1977, p. 123–127.

Bonnefille, R., Lobreau, D., and Riollet, G., 1982, Pollen fossile de *Ximenia* (Olacaceae) dans le Pléistocène Inférieur d'Olduvai en Tanzanie; Implications paléoécologiques: Journal of Biogeography, v. 9, p. 469–486.

Cerling, T. E., 1984, The stable isotopic composition of soil carbonate and its relationship to climate: Earth and Planetary Science Letters, v. 71, p. 229–240.

Cerling, T. E., and Hay, R. L., 1986, An isotopic study of paleosol carbonates from Olduvain Gorge: Quaternary Research, v. 25, p. 63–78.

Cerling, T. E., Hay, R. L., and O'Neil, J. R., 1977, Isotopic evidence for dramatic climatic changes in East Africa during the Pleistocene: Nature, v. 267, p. 137–138.

Cooke, H.B.S., 1957, The problem of glacio-pluvial correlation in east and southern Africa: Proceedings of the 3rd Pan-African Congress on Prehistory, Livingstone, 1955, p. 51–55.

——, 1958, Observations relating to Quaternary environments in East and southern Africa: Geological Society of South Africa Transactions, annexure to v. 60, p. 1–73.

Flint, R. F., 1959, On the basis of Pleistocene correlation in East Africa: Geological Magazine, v. 96, p. 265–284.

Hay, R. L., 1963a, Stratigraphy of Beds I through IV, Olduvai Gorge, Tanganyika: Science, v. 139, p. 829–833.

——, 1963b, Zeolitic weathering in Olduvai Gorge, Tanganyika: Geological Society of America Bulletin, v. 74, p. 1281–1286.

——, 1966, Zeolites and zeolitic reactions in sedimentary rocks: Geological Society of America Special paper 85, 130 p.

——, 1967, Revised stratigraphy of Olduvai Gorge, *in* Bishop, W. W., and Clark, J. D., eds., Background to evolution in Africa: Chicago, Illinois, University of Chicago Press, p. 221–228.

——, 1968, Chert and its sodium-silicate precursors in sodium-carbonate lakes in East Africa: Contributions to Mineralogy and Petrology, v. 17, p. 255–274.

——, 1970, Silicate reactions in three lithofacies of a semiarid basin, Olduvai Gorge, Tanzania, *in* Morgan, B. A., ed., Fiftieth Anniversary Symposia; Mineralogy and petrology of the upper mantle, sulfides, and mineralogy and geochemistry of non-marine evaporites: Mineralogical Society of America Special Paper 3, p. 237–255.

——, 1971, Geologic background of Beds I and II, *in* Leakey, M. D., Olduvai Gorge, Volume 3: Cambridge University Press, p. 9–18.

——, 1973, Lithofacies and environments of Bed I, Olduvai Gorge, Tanzania: Quaternary Research, v. 3, p. 541–560.

——, 1976, Geology of the Olduvai Gorge: University of California Press, 203 p.

Hay, R. L., and Moiola, R. J., 1963, Authigenic silicate minerals in Searles Lake, California: Sedimentology, v. 2, p. 312–332.

Hopwood, A. T., 1951, The Olduvai fauna, *in* Leakey, L.S.B., Olduvai Gorge: Cambridge University Press, p. 20–24.

Johanson, D. C., and 9 others, 1987, New partial skeleton of *Homo habilis* from Olduvai Gorge, Tanzania: Nature, v. 327, p. 205–209.

Kappelman, J., 1984, Plio-Pleistocene environments of Bed I and lower Bed II, Olduvai Gorge, Tanzania: Palaeogeography, Palaeoclimatology, Palaeoecology, v. 48, p. 171–196.

——, 1986, Plio-Pleistocene marine-continental correlation using habitat indicators from Olduvai Gorge, Tanzania: Quaternary Research, v. 25, p. 141–149.

Kleindienst, M. R., 1964, Summary report on excavations at site JK 2, Olduvai Gorge, Tanganyika, 1961–1962, Annual Report of the Antiquities Division of Tanganyika, Appendix 2, p. 4–6.

——, 1973, Excavations at Site JK 2, Olduvai Gorge, Tanzania, 1961–1962; The geological setting: Quaternaria, v. 17, p. 145–208.

Krynine, P. D., 1950, Petrology, stratigraphy, and origin of the Triassic sedimentary rocks of Connecticut: Connecticut Geology and Natural History Survey Bulletin 73, 247 p.

Leakey, L.S.B., 1931, The stone age cultures of Kenya colony: Cambridge University Press, 288 p.

——, 1951, Olduvai Gorge: Cambridge University Press, 164 p.

——, 1957, Preliminary report on a Chellean I living site at BK II, Olduvai: Proceedings of the 3rd Pan-African Congress on Prehistory, Livingstone, 1955, p. 217–218.

——, 1965, Olduvai Gorge, 1951–1961: Cambridge University Press, 118 p.

——, 1969, Age of Bed V, Olduvai Gorge, Tanzania: Science, v. 166, p. 532.

Leakey, L.S.B., and Solomon, J. D., 1929, East African archeology: Nature, v. 124, p. 9.

Leakey, L.S.B., Evernden, J. F., and Curtis, G. H., 1961, The age of Bed I, Olduvai Gorge, Tanganyika: Nature, v. 191, p. 478–479.

Leakey, M. D., 1967, Preliminary survey of the cultural material from Beds I and II, Olduvai Gorge, Tanzania, *in* Bishop, W. W., and Clark, J. D., eds., Background to evolution in Africa: Chicago, Illinois, University of Chicago Press, p. 417–446.

——, 1971, Olduvai Gorge, Volume 3: Cambridge University Press, 306 p.

——, 1975, Cultural patterns in the Olduvai sequence, *in* Butzer, K. W., and Isaac, G. L., eds., After the australopithecines: Paris, Mouton Publishers, p. 477–494.

——, 1978, Olduvai fossil hominids; Their stratigraphic positions and associations, *in* Jolly, C. J., ed., Early African hominids: London, Duckworth, p. 3–16.

Leakey, M. D., and Hay, R. L., 1982, The chronological position of the fossil hominids of Tanzania, *in* L'Homo erectus et la place de L'Homme de Tautavel parmi les Hominidés Fossiles: Proceedings of the First Congress of Human Paleontology, Nice, 1982, v. 2, p. 753–765.

Pickering, R., 1960, A preliminary note on the Quaternary geology of Tanganyika, *in* Joint meeting, Leopoldville, 1958, of east-central, west-central, and southern committees for geology: Commission for Technical Cooperation in Africa South of the Sahara Publication 44, p. 77–89.

——, 1962, A contribution to the geology of the Olduvai Gorge: Geological Society of Tanganyika Report RP/23 (unpublished).

Potts, R., 1984, Home bases and early hominids: American Scientist, v. 72, p. 338–347.

Reck, H., 1914, Erste vorläufige Mitteilung über den Fund eines fossilen Menschenskelets aus Zentralafrika: Sitzungsbericht der Gesellschaft naturforschender Freunde zu Berlin, no. 3, p. 83–95.

——, 1951, A preliminary survey of the tectonics and stratigraphy of Olduvai, *in* Leakey, L.S.B., Olduvai Gorge: Cambridge University Press, p. 5–19.

Smith, G. I., 1984, Paleohydrologic regimes in the southwestern Great Basin, 0–3.2 m.y. ago, compared with other long records of "global" climate: Quaternary Research, v. 22, p. 1–17.

Speth, J. D., and Davis, D. D., 1976, Seasonal variability in early hominid predation: Science, v. 192, p. 441–445.

Walker, T. R., 1967, Formation of red beds in modern and ancient deserts: Geological Society of America Bulletin, v. 78, p. 353–368.

Wayland, E. J., 1932, The Oldoway Report: mimeographed report dated September 2, 1932, 13 p.

MANUSCRIPT ACCEPTED BY THE SOCIETY JUNE 6, 1989

Printed in U.S.A.

Geological Society of America
Special Paper 242
1990

The establishment of a chronological framework for the hominid-bearing deposits of Java; A historical survey

B. Theunissen
Institute for the History of Science, Utrecht University, Nieuwe Gracht 187, 3512 LM Utrecht, The Netherlands
J. de Vos
Rijksmuseum van Natuurlijke Historie, P.O. Box 9517, 2300 RA Leiden, The Netherlands
P. Y. Sondaar
Institute for Earth Sciences, Utrecht University, P.O. Box 80.021, 3508 TA Utrecht, The Netherlands
F. Aziz
Geological Research and Development Centre, Jalan Diponegoro 57, Bandung, Indonesia

INTRODUCTION

There is a long-standing tradition of paleontological and paleoanthropological research in Java. The first fossil mammals were discovered in the middle of the nineteenth century, the first hominids came to light in the 1890s, followed by many more in this century, and research continues.

In the years 1891 to 1893, E. Dubois discovered the famous remains of *Pithecanthropus* (now *Homo*) *erectus* near Trinil; this was the first hominid ever to be accepted as evidence for human evolution. During excavations by the Geological Survey in 1931 to 1933, W.F.F. Oppenoorth found eleven skulls at Ngandong, which he described as *Homo soloensis.* More remains of *Pithecanthropus* and other fossil hominids were discovered by G.H.R. von Koenigswald near Mojokerto and Sangiran in the years 1936 to 1940. Since then, more than 30 finds of hominid fossils have come to light in the Sangiran area. An important additional find was made at Sambunmacan in 1973.

The age of the Javanese hominids has been disputed from the beginning, and there are still lively discussions on the subject. The age problem is central to much of the recent research in Java.

In this paper we present a historical survey of the discovery of the most important hominid-bearing deposits in Java, and of the attempts to establish a chronological framework for the Javanese hominids.

DISCOVERY OF THE HOMINID-BEARING DEPOSITS

The deposits in which the Javanese fossil hominids were to be found became known in the second half of the nineteenth century. The first vertebrate and mammal fossils were discovered accidentally by the local population and by roving naturalists. Some of these naturalists actively searched for fossils, but none of them carried out systematic excavations.

In his account of the natural history of Java the well-known naturalist and traveller Franz Junghuhn (1853) mentioned only one fossil vertebrate: the shark, *Carcharias megalodon.* Although he had searched for mammal fossils in several caves in Java's Tertiary mountains, he had not succeeded in finding any. In 1857, however, after a second visit to the island, he reported a rich site with extinct proboscideans on Mount Pati Ajam in the Japara residency (Fig. 1). He described three fossil elephants from this "Battlefield of Giants," as he called it: *Elephas primigenius,* which was the most abundant, *Mastodon elephantoides,* and *Elephas* sp., which differed from all known extinct and extant elephants. Besides these pachyderms, he also mentioned a dubious bovine species (*Bos*).

On the basis of the presence of *Mastodon elephantoides,* the Dutch geologist W.C.H. Staring concluded in 1865 that the fauna to which it had belonged was of "diluvial" (Pleistocene) age.

In 1859 the Dutch baron A. Sloet van Oldruitenborgh described some fossils that he had collected near Dessa Teguan in the vicinity of Pandan in the Kendeng Hills. P.E.C. Schmulling (1864) reported on some fossil mammals from the Kaleoso district (Dessa Sangiran), and the Javanese painter and naturalist Radèn Saléh (1867) described fossils he collected in 1866 and 1867 on Mount Pandan and in Kedung Lumbu (near Kedung Brubus) in the Kendeng Hills.

The collections of Junghuhn, Radèn Saléh, and most other collectors were sent to the National Museum of Geology and

Theunissen, B., de Vos, J., Sondaar, P. Y., and Aziz, F., 1990, The establishment of a chronological framework for the hominid-bearing deposits of Java; A historical survey, *in* Laporte, L. F., ed., 1990, Establishment of a geologic framework for paleoanthropology: Boulder, Colorado, Geological Society of America Special Paper 242.

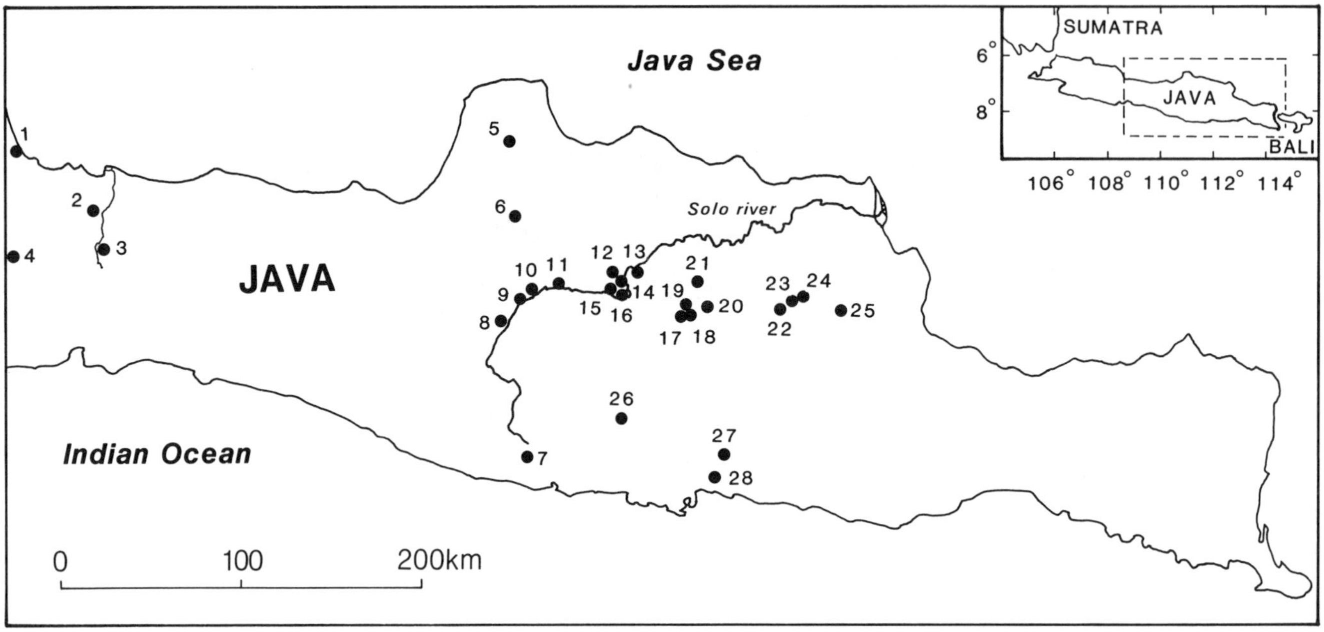

1 Cheribon	5 Pati Ajam	9 Gesi	13 Tinggang	17 Notopuro	21 Bareng	25 Mojokerto
2 Limbangan	6 Ngembak	10 Baringinan	14 Ngandong	18 Butak	22 Kalibeng	26 Sampung
3 Bumiaju	7 Punung	11 Sambunmacan	15 Trinil	19 Kedung Brubus	23 Kabuh	27 Tulung Agung
4 Tji Djulang	8 Sangiran	12 Randublatung	16 Watualang	20 Pandan	24 Pucangan	28 Wajak

Figure 1. Java. The figure shows all important localities mentioned in the text.

Mineralogy in Leiden, the Netherlands. The geologist Karl Martin, director of the museum, described them in the 1880s. His analysis of the fossils led Martin to the conclusion that the fossil mammal fauna of the Kendeng Hills, Mount Pati Ajam, and Ngembak was equivalent to the Tertiary fossil fauna of the Siwalik Hills and of Narbada in British India, described by the British paleontologists H. Falconer, P. T. Cautley, and R. Lydekker (Martin, 1883, 1884, 1887, 1888, 1890). Martin listed the following species: the proboscideans *Stegodon trigonocephalus* n.sp., *Stegodon bombifrons* (Siwaliks), *Stegodon cliftii* (Siwaliks), *Euelephas namadicus* (Narbada), and *Euelephas hysudricus* (Siwaliks); the suid *Sus hysudricus* (Siwaliks); *Bison sivalensis* (?) (Siwaliks); the bovid *Bos* sp.; and the deer *Cervus lydekkeri* n.sp.

The first hominid remains were found in the Kendeng Hills in 1890 by Eugène Dubois.

EUGÈNE DUBOIS; THE DISCOVERY OF PITHECANTHROPUS

Marie Eugène François Thomas Dubois (Fig. 2) was born in 1858, a year before Darwin published his *Origin of Species,* and two years after the Neanderthal skull had been found in the Neander Valley near Düsseldorf. He grew up in a period that witnessed the rapid acceptance and dissemination of the theory of evolution. In the 1860s and 1870s the problem of human ancestry was central to many discussions on evolutionary theory. Until far into the 1880s, opponents and adherents of an evolutionary ancestry for humans were agreed that no hominid fossils were known that provided proof of human evolution. The Neanderthal skull was considered by some as the skull of an idiot, an "old Dutchman," or a Cossack. Others regarded it as belonging to a member of a primitive human race. It was Huxley (1863) who set the tone of the discussion by describing the morphological characteristics of the Neanderthal as primitive, yet definitely human. He also pointed to the large brain capacity as proof of the Neanderthal's human nature. Later finds, such as the mandible of La Naulette (1866) and the Spy skeletons (1887), were also generally ascribed to a (primitive) race of modern humans (Erickson, 1976).

Dubois was among the first to bring about a change in this climate of opinion (Theunissen, 1989). Born in Eijsden in the south of the Netherlands, near St. Peters' Mountain where the remains of a mosasaur had been found, Dubois had been interested in paleontology from his early childhood. His interest in paleoanthropology was probably encouraged by the German anthropologist Carl Vogt, who in the late 1860s lectured in the Netherlands on evolutionary theory and human descent. From 1877 to 1884 Dubois studied medicine at the University of Amsterdam, where he became Reader in anatomy in 1886. Although he now seemed to be on the brink of a prosperous academic career, his predilection for paleoanthropology made him decide to change his course: more and more he became possessed by the

Figure 2. Eugène Dubois in 1928.

idea of beginning a search for hominid fossils that might prove human evolution. Following this call, he gave up his position at the university and, in 1887, left for the Dutch East Indies (Indonesia) to begin his search.

Why Indonesia? Dubois (1889) referred to Darwin, Wallace, and Lydekker to explain this choice. In his *Descent of Man,* Darwin (1871) had reasoned that our distant human ancestors must have lived in the tropics, since human beings had lost their fur pelts in the course of their development. He had suggested Africa, where the chimpanzee and the gorilla live, as the most probable region of human origins. Wallace, on the other hand, had stressed the importance of searching for the ancestors of the present-day anthropoids in caves and Tertiary deposits in both Africa and Southeast Asia. Finally Richard Lydekker (1879) had described a primate fossil—an incomplete upper jaw with a number of teeth—from the Siwalik Hills in British India, which seemed to throw some light on human descent. According to Lydekker, this primate, which he named *Palaeopithecus sivalensis,* could be regarded as a predecessor of the chimpanzee. Yet, he added that the fossil also showed resemblances to both gibbon and human. Dubois concluded from this that "the Hylobates group which in earlier geological periods had developed further" may have played a role in human evolution (1889, p. 160–161).

For Dubois the East Indies thus seemed a suitable area, the more so because this colony of the Netherlands lay wide open to him. In order to maintain himself he joined the Dutch East Indies Army as a medical officer, and in December 1887 he arrived in Padang on Sumatra. In May 1888 he was seconded to Pajakombo in the Padang Highlands, where he had the opportunity to begin a search for caves suited for paleontological excavation. By August he had found some proof that fossil mammals were actually to be found in the caves of Sumatra. Thereupon the colonial government enabled Dubois to dedicate all his time to his search and charged him to carry out paleontological excavations in Sumatra and, if necessary, in Java. Two members of the army engineering corps were assigned to his party along with 50 forced laborers to help him with the excavation work.

The Sumatran cave fauna soon proved too young to include any human forerunners. In 1890 Dubois therefore decided to continue his excavations in Java, where fossils of a supposedly Tertiary age had already been found. Moreover, in 1888, mining engineer B. D. van Rietschoten had found a fossil skull near the village of Wajak in Java, which though fully human, clearly differed from the modern Javanese population (Dubois, 1890). Java thus held promising prospects.

Cave exploration in Java proved unrewarding, and Dubois now turned his attention to the open field. Success was almost immediate. In November 1890, near Kedung Lumbu at Kedung Brubus in the Kendeng Hills, he found a fragment of a mandible that he described as follows: "Amidst the remains of typical representatives of the fauna concerned, and in the same layer . . . a human fossil was found, the right side of the chin of a lower jaw with the sockets of the canine tooth and of the first and second premolar. . . . [T]his fossil [jaw] forms a different and probably lower type than any previously known" (1891, p. 14–15).

In August 1891, during the second digging season in Java, Dubois began excavations at Trinil, a locality that was to acquire historical significance. An enormous number of vertebrate fossils were unearthed, and in September, the first remains of a primate, a third molar, emerged from the sediments. At first, Dubois ascribed the fossil to a chimpanzee (*Anthropopithecus*). In one of his reports to his superiors he wrote: "This genus of anthropoids, occurring only in West- and Central-equatorial Africa today, lived in British India in the Pliocene and, as we can see from this discovery, during the Pleistocene in Java" (1892a, p. 13–14).

The famous skullcap was unearthed in October. In Dubois' opinion it was clearly distinguishable from that of the orang-utan and the gorilla. It had to belong to the same fossil chimpanzee from which a molar had been found a month earlier. Despite having designated the find as a "chimpanzee," Dubois was well aware that he had made an important discovery. The Javanese cranium was higher and larger than that of the recent chimpanzee and substantially more human-like than any known anthropoid, whether fossil or recent (1892b, p. 14–15).

In August of the following year a third primate fossil was

Figure 3. The site at Trinil in 1900.

discovered, this time an almost completely preserved left femur. "This thigh bone," Dubois stated, "lay at the same level in which both the other parts were found, yet following the direction of the earlier stream which deposited the tufa material 15 m upstream. From the circumstances of the find and [my] comparative research it is evident that the three skeletal elements belong to one and the same individual, probably a very aged female" (1893, p. 10). The almost perfectly human characteristics of the femur indicated that the individual must have walked upright, and this induced Dubois to christen his "chimpanzee" *Anthropopithecus erectus.* Further he stated: "In view of all three skeletal elements, especially the femur, *Anthropopithecus erectus* Eug. Dubois approaches modern Man more closely than any of the three great apes, a fact which is in harmony with the thesis of Lamarck, and later of Darwin and others, that the first step in the direction of humanization of our ancestors was the [acquisition of the] erect position."

Additional investigation of the remains finally convinced Dubois that they represented an intermediary form between humans and apes. He therefore decided that it was more appropriate to designate his find an "ape-man," a *Pithecanthropus,* instead of a "man-ape," an *Anthropopithecus.* (The name *Pithecanthropus* was coined by Ernst Haeckel [1868] for the then still hypothetical link between humans and fossil apes.) In 1894 Dubois published the results of his studies under the title *Pithecanthropus erectus, eine menschenaehnliche Uebergangsform aus Java.*

In 1895 Dubois returned to the Netherlands. (The excavations were continued in a systematic way by his assistants until 1900; see Fig. 3). He widely publicized his *Pithecanthropus* finds and displayed them at several international congresses and scientific meetings. Opinions on his discovery varied widely, but essentially the critics were divided into two camps. Some gave the *Pithecanthropus* remains the same treatment the Neanderthal fossils had received 40 years earlier: they regarded them as primitive, though fully human. Others, however, ascribed the bones to an (upright-walking) ape. Dubois (1896) did not fail to exploit this difference of opinion. He pointed out that for some of his colleagues the fossils were apparently too primitive to be regarded as human, while for others they were too human-like to be assigned to an ape. Consequently, *Pithecanthropus* must have been something in between.

After Dubois had clarified several points, and especially after they had been able to examine the fossils for themselves, a growing number of scientists recognized that they were dealing with a transitional form linking humans with their ape-like ancestors. Thus, they accepted Dubois' belief that a phylogenetic significance must be ascribed to the fossils. Most of them did not agree with Dubois' contention that *Pithecanthropus* stood exactly halfway between human and ape. They relegated him to an extinct side branch of human evolution, for instance, or they thought that he had been much closer to *Homo sapiens* than Dubois was prepared to allow. The crucial factor is, however, that they showed themselves ready to adopt an evolutionist interpretation, meaning that for the first time a group of researchers acknowledged a fossil hominid as a transitional form (Theunissen, 1989).

In 1907 and 1908 Dubois published short descriptions of the new species he had discovered, and a summary discussion of the age of the fauna. He considered the fossils discovered in several localities in the Kendeng Hills (such as Kedung Brubus, Teguan, Trinil, and Pati Ajam) to belong to one fauna, the Kendeng or Trinil fauna. He rejected the view, expressed by Martin (1888), that the fauna was identical to the Siwalik Hills fauna in

British India, arguing that the Javanese fauna was younger and probably dated from the late Pliocene. He did note some differences in the faunal compositions of Trinil and Kedung Brubus in that Trinil seemed to be poorer in species. In his view this had to be ascribed to differences in the paleoenvironment. Dubois gave no detailed treatment of the geology of the Trinil layers.

THE SELENKA EXPEDITION

The early finds of hominid fossils, such as those of the Neanderthals, had all been accidental finds. Dubois was the first to undertake a deliberate search for human ancestors. In this way he made an important contribution to the emergence of paleoanthropology as a professional scientific discipline (Theunissen, 1989). An ever-increasing number of scientists set out to follow in Dubois' footsteps; one of the first among these was Lenore Selenka.

The German zoologist Emil Selenka, who had been professor of zoology in Leiden from 1868 to 1874, visited Dubois in 1899 to study the *Pithecanthropus* remains. He was fascinated, and conceived a plan for continuing Dubois' work in Java. However, he died in 1902 before he could carry out this plan. His wife Lenore then decided to take over her husband's work. In 1906 she organized an expedition to Java, which until 1908 carried out excavations near Trinil, close to the spot where Dubois had found his *Pithecanthropus.*

The Selenka expedition did not succeed in finding new remains of *Pithecanthropus,* but a vast collection of fossil animal remains was unearthed. While Dubois had worked alone, Selenka assembled an interdisciplinary team for her excavations and for the publication of the results. Detailed reports were prepared by the geologists W.F.F. Oppenoorth, E. Carthaus, and C. Dozy, and by the paleontologists J. Felix (Anthozoa), H. Martin-Icke (Gastropoda), K. Martin (fresh-water molluscs), E. Hennig (fish), W. Janensch (reptiles and Proboscidea), H. Stremme (mammals, excluding the elephants), and J. Schuster (flora of the Trinil layers). These reports were published in Selenka and Blanckenhorn (1911).

Dubois and Selenka had excavated at almost the same sites, Selenka concentrating on the right bank of the Solo River, Dubois on the left, and they had explored the same fossil-rich layer (the so-called *Hauptknochenschicht*). Contrary to what is commonly believed, Dubois' excavations had been of wider scope than Selenka's. He excavated for 49 months and covered an area of 2,423 m^2, while Selenka worked for 11 months, covering an area of 1,060 m^2 (de Vos and Aziz, 1989). Yet Dubois never settled down to publicize his results extensively (apart from the *Pithecanthropus* remains), so that the Selenka and Blanckenhorn publication constitutes the first detailed description of the geology and of the fossil fauna of Trinil.

Stremme (1911 a, b) and Janensch (1911) gave full descriptions of the mammalian faunal elements discovered by the Selenka expedition. Stremme compared their faunal list with that of Dubois (Fig. 4). Like Dubois, Stremme (1911a, b) and Blanckenhorn (1911) concluded that the Kendeng fauna, including the Trinil finds, was of late Pliocene age.

Dubois Collection	Selenka Collection
Rodentia:	
Hystrix sp.	*Hystrix* sp.
Edentata:	
Manis palaeojavanica	
Carnivora:	
	Mececyon trinilensis
Felis oxygnatha	
Felis trinilensis	
Felis microgale	
	Feliopsis palaeojavanica
Hyaena bathygnatha	
Lutra palaeoleptonyx	
Ungulata:	
Stegodon ganesa var. *javanicas*	*Stegodon airawana*
	[*Stegodon* cf. *trigonocephalus*]
Elephas hysudrindicus	[*Elephas* sp. cf. *antiquus*]
Rhinoceros sivasondaicus	*Rhinoceros sivasondaicus*
Rhinoccros kendengindicus	
Tapirus pandanicus	
Sus brachygnathus	*Sus brachygnathus*
Sus macrognathus	*Sus macrognathus*
Hippopotamus (Hexaprotodon) sivajavanicus	*Hippopotamus* sp.
Cervulus sp.	*Cervulus kendengensis*
Cervus (Axis) liriocerus	*Cervus (Axis) lydekkeri*
Cervus (Rusa) kendengensis	[*Cervus* sp.]
Cervus (Rusa) palaeomendjangan	
Tetraceros kroesenii	*Duboisia kroesenii*
Leptobos groeneveldtii	
Leptobos dependicornis	
Bibos palaeosondaicus	*Bibos palaeosondaicus*
Bibos protocavifrons	
Bubalus palaeokerabau	*Buffelus palaeokerabau*
Primates:	
Macacus or *Semnopithecus*	*Macacus* sp.
Pithecanthropus erectus	

Figure 4. Comparison of the faunal list of the Dubois Collection and the Selenka Collection. The faunal list of Dubois is composed of the faunas of the Kendeng localities, while the Selenka list consists of fossil mammals of the Trinil locality sensu stricto (after Stremme, 1911a, b).

THE GEOLOGICAL SURVEY

In the late 1920s and 1930s, when prospecting for oil and mineral resources became more and more important, scientists employed by the Dutch East Indies Geological Survey—later called "Opsporingsdienst" (i.e., Prospecting Service)—took the lead in the geological and paleontological investigation of Java. Many new sites were explored, important collections were assembled, and eventually new remains of *Pithecanthropus* came to light.

The Dutch mining engineer and employee of the Geological Survey, L.J.C. van Es, became interested in the *Pithecanthropus* question when, in 1926, he made a tour of the island with the paleontologist William D. Matthew of the American Museum of Natural History. He became especially attracted to the still-unsolved problem of the age of *Pithecanthropus erectus,* and decided to devote his Ph.D. research to this subject. In his thesis, which was published in 1931 and dedicated to Eugène Dubois, van Es gave a survey of the literature on the subject and concluded that knowledge of the geology of the "Trinil beds" still left much to be desired. The results of his own investigation of several sites led him to consider the Trinil layers south of the Kendeng Hills from Gesi to Trinil as of the same age as the vertebrate beds at Mount Pati Ajam and the main beds at Sangiran and Baringinan; also of this age were the beds in the northern part of the Kendeng, where vertebrates had been described from Ngembak Hill, Randublatung, Tinggang, Tjapu, and Bareng. He based these conclusions on the vertebrate remains present in the localities. On the basis of the percentage of extant species of fresh-water molluscs in the fauna (83 percent) van Es reached the same conclusion as Martin-Icke (1911), that the vertebrates from all these localities must be of Pleistocene age. Younger remains were to be found on the banks of the Tjitarum River; at these sites, already mentioned by Stehn and Umbgrove (1929), there were no fossils of *Stegodon, Elephas,* and *Hippopotamus.*

Although no new remains of *Pithecanthropus* were found in the 1920s, the Geological Survey did excavate a great many vertebrate fossils, partly from the already well-known localities, partly from a new find-spot in the vicinity of Bumiaju in central Java. The fossils were collected in a systematic way and were sent to the Netherlands for description by the geologist and paleontologist F. H. van der Maarel. In his doctoral thesis, van der Maarel (1932) concluded that the Bumiaju fauna, which contained *Mastodon,* was older than the Trinil fauna. As the Trinil fauna had to be of early or, at the latest, middle Pleistocene age, the Bumiaju fauna was in all probability of late Pliocene or early Pleistocene age.

In the same year, the Dutch geologist W.F.F. Oppenoorth, who had participated in the Selenka expedition back in 1906 to 1908 and who was now employed by the Geological Survey, described new hominid finds from the village of Ngandong on the banks of the Solo River. The site, a river terrace some 20 m above sea level, had already been mentioned by Elbert (1907), and turned out to be a rich deposit of fossil vertebrates, including eleven skulls of a new hominid. Oppenoorth (1932) subsequently described this hominid as *Homo soloensis.* According to Elbert (1907, 1908) the terrace was no older than late Pleistocene. The "Trinil beds" were older, in his view, but younger than Pliocene. Oppenoorth considered the Ngandong fauna as Pleistocene and referred it to the Riss-Würm interglacial. He suggested that a careful investigation of the fauna from Ngandong and a comparison with the other Javanese faunas might provide a more detailed division of the Javanese Pleistocene. It was von Koenigswald who would adopt this paleontological approach.

Gustav Heinrich Ralph von Koenigswald (1902 to 1982) was of tremendous importance to paleoanthropological and biostratigraphical research in Java. Son of the well-known ethnologist G. A. von Koenigswald, he was born in Berlin on 13 November 1902. In 1911 he befriended Kurt Martin, son of the anthropologist Rudolf Martin, who had been intensely involved in the discussions on Dubois' *Pithecanthropus* finds (Martin, 1895, 1896). It was through Rudolf Martin, as von Koenigswald later acknowledged, that he became interested in paleoanthropology (Marks and Drooger, 1983; Tobias, 1984).

Von Koenigswald studied geology and paleontology at several German universities. In his student days his interest in paleoanthropology may have been stimulated by his familiarity with the Haberer Collection of "dragon bones." These bones were brought to the attention of paleontologists and anthropologists by K. A. Haberer, a German naturalist who traveled to China in 1899, hoping to explore the Chinese hinterland. However, the disturbances caused by the Boxer Rebellion confined him to the ports. Here he explored the many drug stores were fossils were sold as medicine. When he returned to Europe, he brought with him a collection of fossils, representing no less than 90 animal species—though not a single reptile, dragon or otherwise, was among them. The collection was described by Max Schlosser in a monograph entitled *Die fossilen Säugethiere China's.* There was only one primate among the remains, represented by an upper molar, which, according to Schlosser, had belonged to a human or an ape. In line with the then-popular conviction that mammalian life had originated in Asia, this prompted him to suggest that human forebears might be found in China.

In 1930, via his professor of paleontology F. Broili, von Koenigswald was offered the post of vertebrate paleontologist of the Dutch Geological Survey in Java, and he gladly accepted this chance to go to Southeast Asia. The first paper he published after his arrival in Java in 1931 dealt with fossils he had purchased from Chinese druggists living on the island (1931a). In the same year he published his first attempt to date the fauna from the "Trinil beds" (1931b).

In a paper published in 1933, von Koenigswald dealt with the fossil vertebrates from Bumiaju. In his view they originated from two layers. The older, the so-called "unterer Wirbeltier Horizont" (the Kali Glagah fauna) was deposited on marine layers with the gastropod *Turitella;* the younger, the "oberer Wirbeltier Horizont," contained a fauna that had been described by Stehlin (1925) as the Limbangan fauna. Also in 1933, von Koenigswald described the fossil vertebrates from the Kendeng Hills and from Ngandong as belonging to two clearly distinct faunas.

Von Koenigswald's official task at the Geological Survey was to provide a division of the terrestrial fluvial deposits of Java on the basis of fossil mammals. In 1934 he established this biostratigraphy. In the introduction to the publication in question he stressed that he would only give a summary of his results and that these had to be regarded as preliminary. Yet von Koenigswald's biostratigraphy would stand for more than 40 years, and it has provided an invaluable guide for all subsequent research in Java. By now, however, it has become clear that it has some serious deficiencies.

Von Koenigswald (1934) based his biostratigraphy on guide fossils and distinguished the following faunas: (recent), Sampung, Ngandong, Trinil, Jetis, Kali Glagah, Tji Djulang (based only on a fragment of a maxillary of the pachyderm *Merycopotamus nanus*), and Tji Sande (based only on a molar of the rhinoceros *Aceratherium boschi*). He used locality names for the divisions of his faunal succession, but his faunal assemblages from the Pleistocene units differed in composition from the fauna actually collected at those "type localities." His Jetis fauna, for instance, was an assemblage of fossils from the north of Jetis and Perning (Cosijn, 1931, 1932), Gunung Butak near Kedung Brubus, and from the lower levels of Sangiran (Sangiran I), whereas his Trinil fauna was an assemblage of fossils from Trinil, Kedung Brubus, the upper levels of Sangiran (Sangiran II) and Pitu (Fig. 5). The difference between the Jetis and Trinil faunas, according to von Koenigswald (1935), was that the bovid *Leptobos* (= *Epileptobos*) and the sabre-toothed cat *Epimachairodus* were present in the Jetis fauna, whereas the specialized *Elephas* ex. aff. *namadicus* was present in the Trinil association.

Besides lumping faunas together, von Koenigswald also changed the faunal lists compiled by previous authors, without producing clear reasons for doing so. He followed Dubois (1908) in stating that Trinil and Kedung Brubus were of the same age. However, his Trinil fauna sensu lato does not include *Tapirus, Manis* (the pangolin), and *Epileptobos,* which are present in Kedung Brubus according to Dubois (1908).

After von Koenigswald had published his biostratigraphy, the Dutch geologist J. Duyfjes, who also worked for the Geological Survey, tried to establish a lithostratigraphy. He studied the geology and stratigraphy of the Javanese Pleistocene mammal-bearing deposits, and published his results in 1936. He proposed the following lithostratigraphy. From old to young: Lower Kalibeng beds, Upper Kalibeng beds, Pucangan beds, Kabuh beds, Notopuro beds and Terrasses. The type localities were several kilometers apart, however.

As was not unusual in those days, Duyfjes attributed a chronostratigraphic value to his lithostratigraphy. The Pucangan Formation and the Kabuh Formation have since been regarded as equivalent in age to the Jetis fauna and the Trinil fauna, respectively (see Fig. 6). Moreover, Duyfjes defined the boundary between the Pucangan and Kabuh Formations on the basis of their fossil content as established by von Koenigswald; lithologically there is no appreciable difference between the two formations. To make things even more confusing, the name "Kabuh beds" had already been used by van Es (1931) to designate a marine marl.

RALPH VON KOENIGSWALD; NEW FINDS OF PITHECANTHROPUS

As a consequence of the economic depression, von Koenigswald lost his position as vertebrate paleontologist for the Geological Survey at the end of 1935. In 1936 he left for the Netherlands to find a new job, but with little success. Consequently, his spirits were low when, in October 1936, he paid a visit to Eugène Dubois in Haarlem. It proved to be a turning point in his career. His enthusiasm for paleoanthropology was revived, and it became clear to him that all he wanted was to return to Java and search for fossil hominids (von Koenigswald, 1956a). After he had arrived there at the beginning of 1937, his financial problems were solved by his appointment as research associate of the Carnegie Foundation. (The support of Teilhard de Chardin, whom he had met in Java in the previous year, was important in this respect because it was Teilhard who introduced him to the foundation.) Now von Koenigswald could devote his full attention to paleoanthropological research.

As early as March 1936 he had been able to announce the find of "the child from Mojokerto," which he was sure belonged to *Pithecanthropus.* Unfortunately, as he stated himself, the skull in question was a surface find and it was rather eroded. Yet on the basis of the guide fossils of the Jetis fauna found in the Mojokerto area, he was convinced that the skull derived from an older layer than Dubois' Trinil finds (von Koenigswald, 1936a).

Dubois, still actively engaged in his *Pithecanthropus* research, had serious doubts about this claim, and he also questioned von Koenigswald's identification of the skull as that of a *Pithecanthropus* (Dubois, 1936a). Soon after this, von Koenigswald (1936b) corrected his description of the circumstances of the Mojokerto find. He now stated that the find was not made at the surface, quoting the geologist J. Duyfjes as his authority: "I [i.e., Duyfjes] can positively assure Prof. Dubois that, at the findspot of the new skull, a mix-up with bones found at the surface is absolutely excluded. [The skull] was excavated by our collector Andojo from a hole, 1m in depth, in hard conglomeratic sandstone, which with certainty belongs to the fossil bone level of Mojokerto." To appease Dubois, von Koenigswald (1936c) changed the name of the fossil from *Pithecanthropus* to *Homo modjokertensis.* Privately, however, he remained convinced that the skull belonged to a *Pithecanthropus* (von Koenigswald, 1956a).

Dubois (1936b) did not accept this new name and attributed the skull to *Homo soloensis* on morphologic grounds. Weidenreich (1939) did the same, yet following von Koenigswald, he dated the Mojokerto find to the early Pleistocene.

Later, von Koenigswald (1956a, p. 79, and 1975, p. 306) would revive the discussion over the age of the skull by stating: "It was found in a shallow excavation by Andojo, one of the Indonesian assistants working for Ir. Duyfjes. . . . Accidentally, this excavation was the only one in that region. Andojo had observed the skull of a *Leptobos* partly exposed by erosion and he excavated the fossil; just underneath it, at a depth of less than one metre, he discovered the human skull." The crucial point here is that *Leptobos* was, in von Koenigswald's biostratigraphy, a representative of the Jetis fauna.

Von Koenigswald made his most important finds at Sangiran. He had been introduced to the area in 1934 by the geologist R. W. van Bemmelen, who later (1968) reminisced that after they had arrived in the area "von Koenigswald settled himself at the foot of the erosion wall of the Pleistocene layers [and] focussed his attention on the dessa boys, who flocked around him, and whom he showed a fossil molar and a fossil bone fragment, exclaiming: 'Onto tjili tjili; Balang kuno' [small molars and giant bones], and telling them how much he would pay for each find. . . . In no time he had sent out a dozen lads who brought him the

	Jetis Fauna			Trinil Fauna				Ngandong Fauna			
	Jetis	Gunung Butak	Sangiran I	Trinil	Kedung Brubus	Sangiran II	Pitu	Ngandong	Kuwukulon	Watualang	Pandejan
Manis palaeojavanica		+									
Hystrix sp.				+		+					
Felis palaeojavanica				+				+		+	+
F. tigris soloensis				+			+	+			
F. cf. *tigris*		+									
F. cf. *pardus*						+		+			
(*F. groeneveldtii*)				+							
(*Hyaena*)				+							
Epimachairodus zwierzyckii n. sp.			+								
Mececyon trinilensis				+							
Cuon crassidens											+
Stegodon trigonocephalus cf. *praecursor*	+	+	+								
St. t. trigonocephalus				+	+	+	+	+	+	+	+
Elephas ex. aff. *namadicus*				?	+	+		+	+	+	
E. cf. *maximus*									+		
E. sp.			+								
Cryptomastodon martini				?		+	+				
Sus macrognathus				+	+	+		+	+	+	
S. brachygnathus				+		+	+	+		+	
S. terhaari							+	+	+	+	
S. vatualangensis							+			+	
S. coerti n. sp.	+		+								
S. cf. *stremmi*			+								
S. ex. aff. *vittatus*								+			
S. sp.		+	+								
Hippopotamus (*Hexaprotodon*) *antiquus* n. sp.	+	+	+								
H. (*Hexaprotodon*) *namadicus*				+	+	+		+	+	+	+
H. sp.			+								
Muntiacus muntjac kendengensis				+							+
M. m. muntjac							+	+	+		+
M. sp.			+								
Cervus (*Axis*) *lydekkeri*	?			+	+	+	+	+		+	
C. (*Axis*) *axis javanicus*							+	+	+	+	+
C. (*Rusa*) *problematicus*	+	+	+								
C. (*Rusa*) *hippelaphus*				+		+		+	+	+	+
C. (*Rusa*) *oppenoorthi*							+				
C. (*Rusa*) cf. *zwaani*	+	+	+								
Duboisia kroesenii	?		?	+	+	+					
Antilope modjokertensis n. sp.	+	+	+								
A. cf. *saätensis* n. sp.			+								
A. sp.	+										
Leptobos cosijni n. sp.	+	+	+								
Bos (*Bubalus*) *bubalis palaeokerabau*				+	+	+	+	+	+	+	+
B. (*Bubalus*) cf. *bubalis*				+	+		+				
B. (*Bibos*) *banteng palaeosondaicus*				+	+	+	+	+	+	+	+
B. (*Bubalus*) sp.	+		+								
B. sp.	+		+								
Tapirus pandanicus			+								
Rhinoceros sondaicus				+		+	+	+	+	+	
Rh. cf. *sondaicus*	+		+								
Macacus sp.				+				+			
Pithecanthropus erectus				+	+	?					
Homo neanderthalensis soloensis								+			

Figure 5. The fossil mammals of the Jetis, Trinil, and Ngandong faunas; the faunas are composed of mammals from different localities (after von Koenigswald, 1934).

Von Koenigswald 1934 BIOSTRATIGRAPHY	Duyfjes 1936 LITHOSTRATIGRAPHY	Indonesian Japanese team local names	SEQUENCES	FISSION TRACK AGE Ma	Semah 1982 BRUNHES-MATUYAMA BOUNDARY	Leinders et al. 1985 BRUNHES-MATUYAMA BOUNDARY Ma	BIOSTRATIGRAPHY
Trinil fauna	Notopuro Formation/beds	Pohjajar F.					
Trinil fauna "Sangiran II"	Kabuh Formation/beds	Bapang Formation	- upper tuff - middle tuff - lower tuff Grenzbank	- 0.71 ± 0.10 0.71 ± 0.09 - 0.78 ± 0.15		Brunhes / Matuyama 0.73	Kedung Brubus fauna Trinil H.K.
Djetis fauna "Sangiran I" "Black Clays"	Pucangan Formation/beds	Sangiran Formation	— T 11 – T 10 — T 9 – T 8 — T 7 – T 6 — T 5 – T 4 — T 3 – T 2 — T 1 – T 0	- 1.16 ± 24 - 1.49 ± 0.32 - 1.51 ± 0.25 top Olduvai 1.67	Brunhes / Matuyama 0.73	base Jaramillo 0.97	Ci Saăt fauna Satir fauna
	Kalibeng Formation/beds	Puren Formation					

Figure 6. Comparison of the different models proposed since 1934 for the stratigraphy of the Javanese hominid-bearing deposits. Included are the dates established by the Indonesian-Japanese team and Semah's and Leinders' interpretation of the magnetostratigraphy of the Sangiran area. (T = tuff.)

fossils that had been washed out of the wall and lay in front of it. Fossils from the Lower Pleistocene Jetis layers had a black patina, and those from the Middle Pleistocene Trinil layers were white, so that it was not difficult to separate them according to stratigraphic level."

Von Koenigswald continued to collect in this way, later helped by Atma, a local purchaser. It was also in this way that he acquired his first *Pithecanthropus* skull from Sangiran, in 1937. It was broken into 40 pieces. Originally the pieces were larger, but the collectors had broken them into smaller ones because they were paid by the piece!

Von Koenigswald reconstructed the skull and sent Dubois a photograph. The latter thought the reconstruction was incorrect, and he levelled a poorly disguised accusation at von Koenigswald that he had deliberately falsified the skull (Dubois, 1938). He soon had to withdraw this accusation, but the relationship between the men, already strained by their discussions over the Mojokerto find, deteriorated rapidly from then on.

Before World War II, von Koenigswald discovered several more fossil hominid remains. Whereas Dubois had thought that the *Pithecanthropus'* position in the human phylogenetic tree was halfway between apes and humans, von Koenigswald (1937) recognized the very humanlike characteristics of the fossils and concluded that they had belonged to a close relative of modern humans. Dubois never accepted this conclusion and continued to oppose von Koenigswald's interpretation of the fossils until his death in 1940. This opposition was inspired by his rather idiosyncratic view of the evolutionary process. In the 1920s and 1930s, after an extensive study of brain weight in vertebrates, Dubois developed a model of saltatory evolution, based on stepwise increase of relative brain weight (see e.g., Dubois, 1930). According to this model, *Pithecanthropus* had originated in a line of apes in which a sudden doubling of the relative brain weight had occurred. Humans, in their turn, had developed from *Pithecanthropus* by another doubling event. Von Koenigswald's new finds, however, indicated that *Pithecanthropus* had been much closer to *Homo sapiens* than Dubois had assumed and thus undermined his model, in which *Pithecanthropus* stood exactly halfway between apes and humans. Yet Dubois never gave up his model, arguing that "his" *Pithecanthropus* was different from von Koenigswald's finds (Dubois, 1940; Theunissen, 1989).

Until recent times many hominid finds have been made at Sangiran. They were attributed to different genera and species, such as *Pithecanthropus erectus, Pithecanthropus dubius, Meganthropus palaeojavanicus,* and so on. Day (1986) listed 34 specimens from Sangiran and three "other finds." The dating of the Sangiran remains has posed serious problems for many years, as most of the fossils were discovered by local collectors and none was excavated under controlled conditions. According to von Koenigswald, the Sangiran area contains a Trinil fauna as well as a Jetis fauna, and thus Kabuh beds and Pucangan beds. The hominid remains he attributed to the Pucangan beds. He always clung to the opinion that the Jetis fauna belongs to the early Pleistocene, the Trinil fauna to the middle Pleistocene, and the Ngandong fauna to the late Pleistocene (1935, 1940, 1956b, 1968, 1975).

This was disputed by Dirk A. Hooijer, curator of the Dubois Collection in Leiden from 1947 to 1979. To Hooijer we owe the first systematically executed taxonomic descriptions of the fossils collected in Java by Dubois, except for the suids and cervids. Hooijer (1951, 1956, 1957, 1962, 1975) accepted the Trinil fauna as middle Pleistocene, but he also considered the Jetis fauna as (earlier) middle Pleistocene. Like von Koenigswald, he based his view on guide fossils. The differences between Jetis and Trinil fauna were small, in Hooijer's opinion; many of the species, like *Leptobos groeneveldtii,* which according to von Koenigswald (1934) were restricted to the Jetis fauna, were also present in the Trinil fauna (Hooijer, 1952, 1975).

In recent times, Indonesian research teams have made important contributions to the understanding of Java's Pliocene and Pleistocene deposits. An important additional *Pithecanthropus* find, a calvarium, was made in Sambunmacan in 1973. According to T. Jacob (1976, 1982), an Indonesian graduate student of von Koenigswald's, it was found in situ in a sandstone layer 4.74 m below the surface in middle Pleistocene Kabuh beds on the south bank of the Solo River. S. Sartono (1979), however, doubted that the Kabuh Formation is present and suggested that the skull came from upper Pleistocene deposits at the site.

RADIOMETRIC DATES

Right from the beginning, researchers attempted to arrange the Javanese hominids in phylogenetic order. The Ngandong skulls were immediately accepted as more advanced and younger than the Trinil finds, and the first skull from Sangiran was always considered to be of the same age as the Trinil skull. Opinions on the other finds varied widely; the age as well as the taxonomy and nomenclature of these specimens are still under discussion. Sartono (1982) suggested the following phylogenetic sequence: From young to old: *Homo erectus ngandongensis, Homo erectus trinilensis, Homo palaeojavanicus robustus, Homo palaeojavanicus sangiranensis.*

Sartono used a morphological approach to arrive at his phylogenetic series. The problem has, from the sixties on, also been tackled by methods of absolute dating. Absolute ages have been reported from various lithologic units and tektites in Java. Many of these ages, however, are contradictory and confusing when they are applied to the dating of various geologic and paleontologic events.

Von Koenigswald (1964, 1968) published two potassium/argon (K/Ar) dates. One was based on a leucite-basalt from the Muriah Volcano in central Java, which was used to date the fauna found in the Muriah complex at Pati Ajam—a typical Trinil fauna with the deer *Axis lydekkeri,* according to von Koenigswald. The age of the basalt was 495,000 ± 100 to 60,000 yr. The second date was based on a tektite from the Trinil layers at Sangiran; it was 730,000 ± 50,000 yr. Von Koenigswald supposed the first date to represent the age of the upper Trinil.

Jacob and Curtis (1971) gave a preliminary K/Ar dating for early humans in Java. A sample from Mojokerto gave a result of 1.9 ± 0.4 Ma. As the fauna of Mojokerto was considered to be of the Jetis kind, Jacob (1972) stated that this date also gives an indication of the antiquity of *Meganthropus* and *Pithecanthropus* from the Jetis beds in Sangiran. According to Curtis (1981), however, the date is of little value due to the very high atmospheric-acid content of the sample. In addition, Matsu'ura (1984) noted that the provenance of the tuff sample is reported differently by the various authors (compare Jacob and Curtis, 1971; Jacob, 1972; Curtis, 1981; and Curtis in Pope, 1983).

A fourth K/Ar date of 1.91 Ma was given by Jacob in 1973 (see Bartstra, 1978, p. 57). This date was obtained from an andesite sample from the Pucangan Formation near Kebonduren, not far from Kedung Brubus.

No consensus was reached on the interpretation of these two K/Ar dates of about 2 Ma for the Jetis beds. Von Koenigswald held the opinion that the date for Mojokerto is "good for the upper Jetis only" (1975, p. 306), but others maintained that the two dates represent maxima, and that the average age of the Jetis vertebrate fauna is closer to 1 Ma (e.g., Isaac and Pilbeam, 1975).

An additional K/Ar date was given by Bartstra (1978). The

sample was collected from a steep slope immediately above the outer bend of the Solo River between the villages of Trinil and Gadjah. Referring to Duyfjes (1936), Bartstra stated that the exposed rocks at that spot belong to the Pucangan Formation. In strong contrast to the two dates mentioned above, the date obtained was 0.5 ± 0.3 Ma.

In 1982, in a survey of the literature, D. W. Orchiston and W. G. Siesser reported that radiometric approaches involving the K/Ar method have been popular in attempts to date the fossil hominid-bearing formations in Java. Yet in their view, most of the available dates are inadequate. Few were reported in detail, some were published without standard deviations, and all of them lacked adequate stratigraphic information.

THE INDONESIAN-JAPANESE TEAM; NEW DATES

In 1975, when both parties became equally convinced of the importance and necessity of such a project, the University of Tokyo and the Geological Survey of Indonesia agreed to plan and implement a joint research project on the stratigraphy, paleontology, and chronology of the Javanese hominid-bearing formations. The project was carried out in the years 1976 to 1979, and the results were published in 1985. A broad spectrum of approaches was followed, including petrological and magnetostratigraphic data; fluorine analyses of bones, antlers, and teeth of fossil mammals; pollen analysis; and fission track; all conducted under controlled stratigraphic conditions. The team concentrated on Sangiran and was the first to carry out large-scale systematic excavations in all the fossil beds present in this area.

Suzuki and others (1985) reported several fission-track dates from zircons in the Kabuh and Pucangan beds (sensu Duyfjes, 1936) in the Sangiran area (Fig. 6):

Kabuh		middle tuff	0.78 ± 0.15 Ma
Pucangan	Black Clays (upper part)	tuff 10	1.16 ± 0.24 Ma
	Black Clays (middle part)	tuff 6	1.49 ± 0.32 Ma
	Black Clays (middle part)	tuff 5	1.51 ± 0.25 Ma

Further, they obtained dates for two tektites, one from an excavation site near where "*Pithecanthropus* VIII," a skull fragment, had been discovered, the other one from a site near the find-spot of "*Pithecanthropus* mandible d." Two dates were determined for each site: for the latter site 0.71 ± 0.09 and 0.70 ± 0.12 Ma, for the former 0.71 ± 0.10 and 0.72 ± 0.13 Ma.

Matsu'ura (1982, 1985) measured the fluorine content of several hominid fossils from Sangiran. These measurements, supplemented by those obtained by Oakley and others (1975), were compared with measurements of the fluorine content of fossil mammal bones from different stratigraphic levels. From his data Matsu'ura concluded that the first presence of humans occurs in the uppermost part of the Black Clays or the Grenzbank.

In 1982 the magnetostratigraphy of the Sangiran area was analyzed by F. Semah from the Laboratoire de Préhistoire of the Muséum National d'Histoire Naturelle in Paris. Semah reported, from top to bottom (see Fig. 6), a normal polarity for the Notopuro Formation, the Kabuh beds, and the Grenzbank. The first reversal of the polarity was found in the uppermost part of the Black Clays, a few meters below the Grenzbank; it was identified as the Brunhes-Matuyama boundary, dated at 0.73 Ma. The next normal polarity was reported from the top of the marine Kalibeng Formation just below the Black Clays and was identified as the top of the Olduvai event, dated at 1.67 Ma, which is compatible with a K/Ar date (2.0 ± 0.6 Ma) from the base of the Black Clays, obtained by Nishimura and others (1981).

Leinders and others (1985), however, have pointed to two problems in this interpretation of the paleomagnetic data. First, the Brunhes-Matuyama boundary of 0.73 Ma is far too low in the section, given the fission-track dates published by the Indonesian-Japanese team. Secondly, the Jaramillo normal period, which spans the period from 0.97 to 0.87 Ma and which should be present in Semah's samples of the Black Clays, is absent in his data. Semah attributed this to incomplete sampling due to landslides. Yet according to Leinders and others (1985), another interpretation is more plausible: the change in polarity from normal to reversed at the top of the Black Clays may not represent the Brunhes-Matuyama boundary, but may correspond to the base of the Jaramillo period (0.97 Ma), which confirms the zircon date of 1.16 ± 0.24 Ma from tuff 10 in the upper part of the Black Clays (Fig. 6). In this situation, too, a paleomagnetic event (the reversed period between the Brunhes and the Jaramillo normal periods) would be missing in Semah's data. It should, however, also be noted that in this model the missing normal polarity is situated in the middle part of the Kabuh beds, that is, in coarse clastic sediment with internal erosional contacts: a hazardous lithology for magnetostratigraphic research.

Biostratigraphic analyses of the Sangiran area were published by Aimi and Aziz (1985). In the lower part of the Black Clays, mainly deer and *Hippopotamus* were found. The bovids *Bubalus* and *Bibos* were reported from the stratigraphically higher Cenglik site. The fauna collected from the still higher Grenzbank proved to be more diverse, including Carnivora, *Stegodon, Rhinoceros, Sus,* and the antelope *Duboisia. Elephas* occurred for the first time in the Kabuh Formation of Sangiran, stratigraphically higher than the Grenzbank.

The joint Indonesian-Japanese team successfully clarified the complicated geologic structure of the Sangiran Dome. On this basis, supplemented by the tracing of a number of pyroclastic markers, they succeeded in establishing "type sections" of Kalibeng, Pucangan, Kabuh, and Notopuro Formations (sensu Duyfjes, 1936). As the original type sections of these formations are located 100 to 150 km east of the Sangiran area, correlation of the Sangiran formations with those in the original type areas might lead to confusion. To avoid this, Itihara and others (1985) proposed new names for the Sangiran area to replace the Kalibeng, Pucangan, Kabuh, and Notopuro Formations; these new names are Puren, Sangiran, Bapang, and Pohjajar Formations, respectively.

NEW IDEAS ON THE FAUNAL SUCCESSION

In several studies of Java's fossil vertebrates, doubts were expressed with regard to the scientific value of the material collected by Dubois. Von Koenigswald (1934), for example, questioned the presence of the giant pangolin, *Manis palaejavanica* at Kedung Brubus; Duyfjes (1936) stated that it is impossible, on the basis of Dubois' data, to distinguish which fossils were collected at Butak and which at Kedung Brubus; Brongersma (1941) expressed the opinion that the collection is valuable from a zoological point of view, but that its geological significance has diminished; and Hooijer (1948) regretted that the fossils were not collected according to modern standards. It was generally believed until the 1970s that the Dubois Collection was insufficiently documented to be used for biostratigraphic investigations.

In the early 1980s, John de Vos, curator of the Dubois Collection since 1979, and Paul Sondaar, from the Utrecht Institute of Earth Sciences, decided to reconstruct Dubois' methods of excavation in order to obtain an up-to-date view of the collection's value. From their investigations, the following picture emerged (de Vos and Sondaar, 1982; de Vos and others, 1982).

During his investigations in Java, Dubois was stationed at Tulung Agung. From here he explored the residencies of Madiun, Kediri, Surakarta, Rembang, and Japara. Excavations were carried out at several places, supervised by his assistants, G. Kriele and A. de Winter, sergeants in the Engineering Corps. Dubois regularly visited the localities to examine the sites, and issue new instructions. The letters in which Kriele and de Winter reported their progress to Dubois contain a great deal of valuable information on the excavations. These letters are now in the Dubois Collection (Rijksmuseum van Natuurlijke Historie, Leiden, The Netherlands). On the basis of these letters, his visits, and his notes, Dubois made monthly progress reports for his superiors. These reports remained unpublished; drafts are present in the Dubois Collection. From these reports, quarterly abstracts were made, which were published in a supplement to the *Javasche Courant*, the *Verslagen van het Mijnwezen;* drafts of these are also preserved in the Dubois Collection. From all these published and unpublished data a good picture can be obtained of how, where, and when Dubois' excavations on Sumatra and Java were carried out.

The way in which Dubois and his assistants did their collecting was sufficiently reliable and accurate; level and exact locality were recorded for the more important specimens. After the fossils had arrived in Leiden, the collection was moved to other locations several times, which has caused some disturbance. The museum was bombed in World War II, yet the greater part of the material and the important specimens remained intact. Most specimens are labeled and catalogued and have good locality data. All this adds up to the conclusion that the Dubois Collection can safely be used for paleofaunistic and biostratigraphic ends.

Having arrived at these conclusions, de Vos and others decided to devise a biostratigraphy of the Kendeng and Trinil area, taking the Dubois Collection as their starting point. They concentrated on Trinil and Kedung Brubus, where fossil hominid remains had been found.

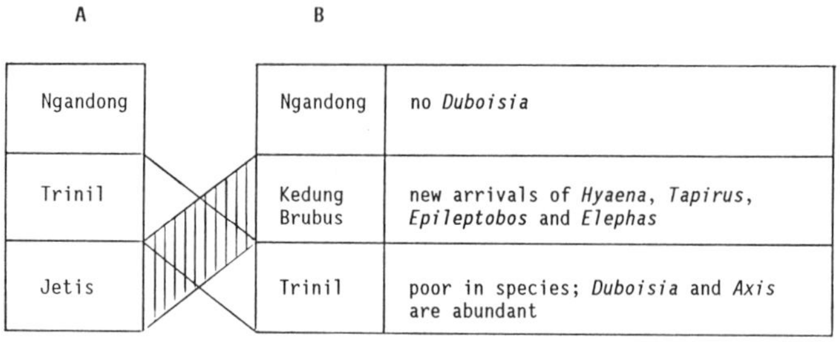

Figure 7. Two views of the faunal succession in Pleistocene Java. A, the view of von Koenigswald (1934); B, the view of de Vos and others (1982). (After de Vos and others, 1982.)

Dubois (1908) had already pointed to differences in faunal composition between Trinil and Kedung Brubus, the fauna from Trinil being poorer in species than that from Kedung Brubus. In his opinion, these differences were due to differences in the paleoenvironment; he considered the age of the faunas to be equivalent. De Vos and others (1982) also found differences between the Trinil and the Kedung Brubus fauna, and they confirmed that the Trinil fauna was much poorer in species. In their view, paleoecological factors alone were insufficient to explain this difference; there must be a difference in age, the Kedung Brubus fauna being the younger because of the presence of the advanced *Elephas hysudrindicus, Rhinoceros kendengindicus, Epileptobos,* and *Tapirus.*

The Jetis fauna (sensu von Koenigswald, 1934) turned out to be similar to the Kedung Brubus fauna, and younger than that from the Trinil locality. On the basis of the relatively low number of species in the Trinil fauna, de Vos and others (1982) suggested that the area was relatively isolated and that there was little faunal interchange with the Asian mainland when the Trinil sediments were deposited, resulting in the presence of endemics like the antelope *Duboisia.* The faunal subdivisions Jetis-Trinil-Ngandong (sensu von Koenigswald, 1934) for the Pleistocene of Java must be abandoned, in their view. Instead, a faunal succession of Trinil-Kedung Brubus-Ngandong should be used, the Trinil fauna being the fauna of the type locality with *Homo erectus* (Fig. 7).

In 1983 de Vos also revised the Punung fauna. This fauna was first included in the Trinil assemblage by von Koenigswald (1939, 1940), but Badoux (1959) separated it from the Trinil fauna and considered it to be late middle Pleistocene, filling the gap between the middle Pleistocene Trinil fauna (sensu von Koenigswald, 1934) and the late Pleistocene and sub-recent Ngandong and Sampung fauna assemblages. According to de Vos, Punung is biostratigraphically intermediate between the Ngandong and the Wajak fauna, the latter described by van den Brink (1982), because the Punung fauna contains the extant *Elephas maximus,* which is absent in Ngandong and has disappeared again in the Wajak fauna. As is evident from the presence of *Pongo pygmaeus*

Faunal succession	Representative taxa	Environment	Some localities with land vertebrates
Recent			
Wajak	*Homo sapiens sapiens* fauna has a recent character with a few extinct species	open woodland	Wäjak
Punung	*Pongo*, *Hylobates*, *Ursus malayanus*, *Elephas maximus*, *Sus vittatus*, *Capricornis*	humid forest	Punung
Ngandong	*Stegodon trigonocephalus*, *Sus macrognathus*, *Elephas hysudrindicus*, *Homo erectus soloensis*	open woodland	Ngandong, Sidorejo
Kedung Brubus	new arrivals of: *Tapirus*, *Hyaena*, *Epileptobos*, *Elephas hysudrindicus*, *Sus macrognathus*, *Stegodon hypsilophus*	open woodland	Kedung Brubus, Jetis, Kabuh of Sangiran
Trinil H.K.	endemics, like *Duboisia* and *Rattus trinilensis*, relatively poor in species. *Meganthropus*, *Rhinoceros*, *Sus brachygnathus*, *Homo erectus erectus*, *Stegodon trigonocephalus*	open woodland	Trinil H.K., Grenzbank of Sangiran
Ci Saät	bovids and large felid		Ci Saät, upper part of the Black Clays of Sangiran, Ci Julang?
Satir	unbalanced island fauna with: *Mastodon*, *Hexaprotodon*, Cervidae and *Geochelone*	partly mangrove forest	Satir (loc. 1-4), middle part of the Black Clays of Sangiran

Figure 8. The proposal of Leinders and others (1985) for the faunal succession of the Javanese hominid-bearing deposits. The model is based on fossil mammals and the giant tortoise *Geochelone.*

(the orang-utan) and of Hylobatidae (gibbons), the Punung fauna was a tropical rainforest fauna.

Sondaar (1984) discussed the faunal succession in more detail, also taking into account the collections made in the 1920s in the Bumiaju area in west Java (van der Maarel, 1932) and those made by the Japanese-Indonesian team in the Sangiran area. The fossils in both collections were excavated in stratigraphic context, and the specimens have good locality data. The lithology of the sites at Bumiaju and Sangiran is different, but the collections are comparable in age and show similar faunal turnovers and changes through time. In broad outline the changes come down to this: a mammal fauna with a low diversity and lacking larger carnivores, which indicates island conditions, is replaced by a mainland fauna in the higher levels. The faunal changes can be interpreted in terms of immigration following the establishment of a better connection with mainland Asia. From old to young, Sondaar arrived at the following series: Satir, Ci Saät, Trinil H.K. (= Hauptknochenschicht), Kedung Brubus, Ngandong, Punung, and Wajak (Fig. 8). He argued that it is unlikely, for paleofaunistic reasons, that humans were present in the unbalanced Satir Island fauna. Thus on different grounds he reached the same conclusion as Matsu'ura (1982)—that no hominids are present in the lower parts of the Black Clays of Sangiran.

In 1985 Leinders and others summarized all data and established the age of the hominid-bearing deposits of Java. They confirmed that the four faunal zones, Satir, Ci Saät, Trinil H.K., and Kedung Brubus can also be identified in the vertebrate-bearing strata around Sangiran (see Fig. 6). Combining the chronostratigraphic dates of the Indonesian-Japanese team with the biostratigraphic model, they arrived at the following ages for the various faunas:

Kedung Brubus	around 0.8 Ma
Trinil H.K.	" 1.0 Ma
Ci Saät	" 1.2 Ma
Satir	" 1.5 Ma

Efforts are now being made to refine Sondaar's scheme (1984) by means of fossil micromammals (rodents), which have been used successfully for biostratigraphic correlation of continental deposits in many parts of the world. Fossil micromammals were for a long time virtually unknown from Java, but in 1983, test sampling by a team from the Geological Research and Development Centre of Indonesia and the University of Utrecht proved their presence in the Plio-Pleistocene deposits. Large-scale sampling and washing (wet screening of the samples) has been completed, and the material is now being analyzed.

SUMMARY AND CONCLUSIONS

In conclusion, we will indicate several important steps in the exploration of the Javanese hominid-bearing deposits, meanwhile summing up our account.

The first systematic excavations in Java were carried out by Dubois in the 1890s. His pioneering efforts directly or indirectly stimulated all later research in Java. Near Trinil, Dubois discovered remains of *Pithecanthropus* (now *Homo*) *erectus,* which were the first hominid fossils ever to be accepted as evidence for human evolution. The vertebrate collection assembled by Dubois is still an invaluable tool for biostratigraphic research, yet he never settled down to describing the collection. The first detailed descriptions of the Trinil fauna and of the geology of the Trinil area were published by the Selenka team that, following in Dubois' footsteps, worked in Java from 1906 to 1908.

In the 1920s and 1930s the Dutch Geological Survey made important contributions toward the establishment of a stratigraphic framework for the hominid-bearing deposits in Java. Systematic excavations were carried out at Bumiaju, and the geology of several sites was described in detail. Von Koenigswald published the first biostratigraphy, which was based on guide fossils; Duyfjes established the first lithostratigraphy. These stratigraphies guided paleontological research in Java until well into the 1970s. However, confusion was caused by an unwarranted mixing of bio-, litho-, and chronostratigraphic data.

The 1930s also witnessed the discovery of hominid skulls at Ngandong by Oppenoorth and of additional *Pithecanthropus* finds by von Koenigswald at Sangiran. The latter finds definitively established *Pithecanthropus erectus* as a separate species and as a forerunner of modern humans. Unfortunately, few of these hominids were excavated under controlled stratigraphic conditions, and many of them were surface finds. The reason for this was that most fossils were obtained from paid local collectors; this was von Koenigswald's customary collecting method. As a consequence, the age and provenance of the Javanese hominids is still uncertain in many cases.

In the years 1976 to 1979, a joint Indonesian-Japanese team made a detailed study of the Sangiran area. Using an unprecedentedly broad spectrum of techniques, they investigated the paleontology, stratigraphy, and chronology of the area, all under controlled stratigraphic conditions. They established dates for several fossil beds and carefully documented the biostratigraphy of the area.

Attempts at dating the Javanese beds were made from the 1960s onward. A variety of approaches was used, such as K/Ar dating and fission track. Later, paleomagnetic analyses were also carried out. The relative age of the several hominid fossils themselves was analyzed by means of the fluorine method. Important advances have by now been made, but there is still a good deal of confusion; a consensus on the age of the hominid-bearing beds has not been reached.

A reinvestigation of the Kendeng and Trinil area was undertaken in the early 1980s by de Vos, Sondaar, and others. First they evaluated the usefulness of the material collected by Dubois for biostratigraphic investigations, on which some doubt had been cast by certain researchers. After reaching a positive conclusion, they used this material to eradicate the confusion caused by the mixing of bio- and lithostratigraphic data in earlier work. This resulted in a revision of the model proposed by von Koenigswald in 1934.

Recently Leinders and others (1985) attempted to combine the results obtained in the last decades in the Sangiran and the Kendeng areas. Their analysis resulted in a proposal for the chronology of the Javanese hominid-bearing deposits. This model is now under discussion; certainly much more research is needed to corroborate it. At the moment the matter is being approached from a new angle: efforts are being made to underpin the model by establishing a biostratigraphy on the basis of micromammals.

ACKNOWLEDGMENTS

The authors wish to express their gratitude to two anonymous reviewers for their helpful comments, to Enid Perlin-West, who corrected the English text, and to Lian Hielkema for her help in many ways.

REFERENCES CITED

Aimi, M., and Aziz, F., 1985, Vertebrate fossils from the Sangiran Dome, Modjokerto, Trinil, and Sambungmacan areas, *in* Watanabe, N., and Kadar, D., eds., Quaternary geology of the hominid fossil bearing formations in Java: Geological Research and Development Centre Special Publication 4, p. 155–197.

Badoux, D. M., 1959, Fossil mammals from two deposits at Punung (Java) [Ph.D. thesis]: Utrecht University, 151 p.

Bartstra, G. J., 1978, The age of the Djetis beds in east and central Java: Antiquity, v. 52, p. 56–58.

Blanckenhorn, M., 1911, Allgemeine Betrachtungen über die wissenschaftlichen Ergebnisse der Selenka-Trinil Expedition, *in* Selenka, M. L., and Blanckenhorn, M., eds., Die *Pithecanthropus*-Schichten auf Java; Geologische und paläontologische Ergebnisse der Trinil Expedition (1907 und 1908): Leipzig, Engelmann, p. 258–268.

Brongersma, L. D., 1941, De verzameling van Indische fossielen (Collectie Dubois): De Indische Gids, v. 63, p. 97–116.

Cosijn, J., 1931, Voorloopige mededeeling omtrent het voorkomen van fossiele beenderen in het heuvelterrein ten Noorden van Djetis en Perning (Java): Verhandelingen van het Geologisch-Mijnbouwkundig Genootschap voor Nederland en Koloniën, Geological Series 9, p. 113–119.

—— , 1932, Tweede mededeeling over het vookomen van fossiele beenderen in het heuvelterrein ten Noorden van Djetis en Perning (Java): Verhandelingen van het Geologisch-Mijnbouwkundig Genootschap voor Nederland en Koloniën, Geological Series 9, p. 135–148.

Curtis, G. H., 1981, Establishing a relevant time scale in anthropological and archaeological research: Philosophical Transactions of the Royal Society, Series B, v. 292, p. 7–20.

Darwin, C., 1859, On the origin of species by means of natural selection, or the preservation of favoured races in the struggle for life: London, John Murray, 490 p.

—— , 1871, The descent of man and selection of relation to sex: London, John Murray, 475 p.

Day, M., 1986, Guide to fossil man, 4th ed.: London, Cassell Ltd., 432 p.

de Vos, J., 1983, The Pongo faunas from Java and their significance for biostratigraphical and paleoecological interpretations: Proceedings Koninklijke Nederlandse Akademie van Wetenschappen, v. 86, p. 417–425.

de Vos, J., and Aziz, F., 1989, The excavations of Dubois (1890–1900), Selenka (1906–1908), and the geological survey of Trinil (Java, Indonesia) by the Indonesian-Japanese team (1976–1977): Geological Research and Development Centre Bulletin (in press).

de Vos, J., and Sondaar, P. Y., 1982, The importance of the Dubois Collection reconsidered: Modern Quaternary Research Southeast Asia, v. 7, p. 35–63.

de Vos, J., Sartono, S., Hardja-Sasmita, S., and Sondaar, P. Y., 1982, The fauna from Trinil, type locality of *Homo erectus;* A reinterpretation: Geologie en Mijnbouw, v. 61, p. 207–211.

Dubois, E., 1889, Over de wenschelijkheid van een onderzoek naar de diluviale fauna van Ned. Indie, in het bijzonder van Sumatra: Natuurkundig Tijdschrift voor Nederlandsch-Indië, v. 48, p. 148–165.

——, 1890, Uit een schrijven van den Heer Dubois te Pajacombo naar aanleiding van een aan dien Heer toegezonden schedel, door den Heer van Rietschoten in zijn marmergroeven in het Kedirische opgegraven: Natuurkundig Tijdschrift voor Nederlandsch Indië, v. 49, p. 209–210.

—— , 1891, Palaeontologische onderzoekingen op Java; Extra bijvoegsel der

Javasche Courant, Verslag van het Mijnwezen over het 4e kwartaal 1890: p. 14–18.

—— , 1892a, Palaeontologische onderzoekingen op Java; Extra bijvoegsel der Javasche Courant, Verslag van het Mijnwezen over het 3de kwartaal 1891: p. 12–14.

—— , 1892b, Palaeontologische onderzoekingen op Java; Extra bijvoegsel der Javasche Courant, Verslag van het Mijnwezen over het 4e kwartaal 1891: p. 12–15.

—— , 1893, Palaeontologische onderzoekingen op Java; Extra bijvoegsel der Javasche Courant, Verslag van het Mijnwezen over het 3de kwartaal 1892: p. 10–14.

—— , 1894, *Pithecanthropus erectus,* einen menschenaenliche Uebergangsform aus Java: Batavia, Landesdruckerei, p. 1–39.

—— , 1896, *Pithecanthropus erectus,* eine menschena ehnliche Uebergangsform: Compte-rendu des seances du troisième congrès international de zoologie, Leyde, 11-16 Septembre, 1895, Leiden, p. 251–271.

—— , 1907, Eenige van Nederlandschen kant verkregen uitkomsten met betrekking tot de kennis der Kendeng-fauna (fauna van Trinil): Tijdschrift Koninklijk Nederlandsch Aardrijkskundig Genootschap, Series 2, v. 24, p. 449–458.

—— , 1908, Das geologische Alter der Kendeng- oder Trinil-Fauna: Tijdschrift Koninklijk Nederlandsch Aardrijkskundig Genootschap, Series 2, v. 25, p. 1235–1270.

—— , 1930, Die phylogenetische Grosshirnzunahme autonome Vervollkommnung der animalen Funktionen: Biologia Generalis, v. 6, p. 247–292.

—— , 1936a, Nieuwe *Pithecanthropus* ontdekt?: Angemeen Handelsblad, 18 April 1936.

—— , 1936b, Racial identity of *Homo soloensis* Oppenoorth (including *Homo modjokertensis* von Koenigswald) and *Sinanthropus pekinensis* Davidson Black: Proceedings Koninklijke Nederlandsche Akademie van Wetenschappen, v. 39, p. 1180–1185.

—— , 1938, On the fossil human skull recently described and attributed to *Pithecanthropus erectus* by G.H.R. von Koenigswald: Proceedings Koninklijke Nederlandsche Akademie van Wetenschappen, v. 41, p. 380–386.

—— , 1940, The fossil human remains discovered in Java by Dr. G.H.R. von Koenigswald and attributed by him to *Pithecanthropus erectus,* in reality remains of *Homo wadjakensis* (syn. *Homo soloensis*): Proceedings Koninklijke Nederlandsche Akademie van Wetenschappen, v. 43, p. 494–496, 842–851, 1268–1275.

Duyfjes, J., 1936, Zur Geologie und Stratigraphie des Kendeng-Gebietes zwischen Trinil und Soerabaja (Java): De Ingenieur in Nederlandsch-Indië, v. 3, section 4, p. 136–149.

Elbert, J., 1907, De nieuwste onderzoekingen over het *Pithecanthropus*-vraagstuk: Natuurkundig Tijdschrift voor Nederlandsch-Indië, v. 67, no. 3-4, p. 125–142.

—— , 1908, Ueber das Alter der Kendeng-Schichten mit *Pithecanthropus erectus* Dubois: Neues Jahrbuch für Geologie und Paläontologie, Abhandlungen, v. 25, p. 648–662.

Erickson, P. A., 1976, Origins of physical anthropology [Ph.D. thesis]: Storrs, University of Connecticut, 348 p.

Haeckel, E., 1868, Natürliche Schöpfungsgeschichte; Gemeinverständliche wissenschaftliche Vorträge über die Entwicklungslehre im Allgemeinen und diejenige von Darwin, Goethe, und Lamarck im Besonderen: Berlin, Reimer, 688 p.

Hooijer, D. A., 1948, Evolutie van zoogdieren in het Quartair van Z. O. Azië: Vakblad voor Biologen, v. 28, no. 7, p. 117–122.

—— , 1951, The geological age of *Pithecanthropus, Meganthropus,* and *Gigantopithecus:* American Journal of Physical Anthropology, v. 9, p. 265–281.

—— , 1952, Fossil mammals and the Plio-Pleistocene boundary in Java: Proceedings Koninklijke Nederlandse Akademie van Wetenschappen, Series B, v. 55, no. 4, p. 436–443.

—— , 1956, The lower boundary of the Pleistocene in Java and the age of *Pithecanthropus:* Quaternaria, v. 3, no. 1, p. 5–10.

—— , 1957, The correlation of fossil mammalian faunas and the Plio-Pleistocene boundary in Java: Proceedings Koninklijke Nederlandse Akademie van Wetenschappen, Series B, v. 60, no. 1, p. 1–10.

—— , 1962, The middle Pleistocene fauna of Java, *in* Kurth, G., ed., Evolution und Hominisation: Stuttgart, G. Fischer, p. 108–111.

—— , 1975, Quaternary mammals west and east of Wallace's line: Netherlands Journal of Zoology, v. 25, no. 1, p. 46–56.

Huxley, T. H., 1963, Man's place in nature, *in* Collected essays, 1893–1894, v. 7: Hildesheim, Georg Olms Verlag, p. 1–208.

Isaac, G. L., and Pilbeam, D., 1975, Correlation charts compilated at the symposium, Appendix 1, *in* Butzer, K. W., and Isaac, G. L., eds., After the Australopithecines; Stratigraphy, ecology and culture change in the middle Pleistocene; Papers of the Burg Wartenstein Symposium, July, 1973: The Hague, Mouton, p. 889–899.

Itihara, M., Kadar, D., and Watanabe, N., 1985, Concluding remarks, *in* Watanabe, N., and Kadar, D., eds., Quaternary geology of the hominid fossil-bearing formations in Java: Geological Research and Development Centre Special Publication 4, p. 365–378.

Jacob, T., 1972, The absolute date of the Jetis beds at Modjokerto: Antiquity, v. 46, p. 148.

—— , 1976, Early populations in the Indonesian region, *in* Kirk, R. L., and Thorne, A. G., eds., The origin of the Australians: Canberra, Australian Institute of Aboriginal Studies, p. 81–93.

—— , 1982, Solo Man and Peking Man, *in* Cybulski, J. S., and Sigmon, B. A., eds., *Homo erectus:* Ontario, University of Toronto Press, p. 87–104.

Jacob, T., and Curtis, G. H., 1971, Preliminary potassium-argon dating of early man in Java: Contributions University of California, Archeological Dating Facility, v. 12, p. 50.

Janensch, W., 1911, Die Proboscidier-Schädel der Trinil-Expeditions-Sammlung, *in* Selenka, M. L., and Blanckenhorn, M., eds., Die *Pithecanthropus*-Schichten auf Java; Geologische und paläontologische Ergebnisse der Trinil-Expedition (1907 and 1980): Leipzig, Engelmann, p. 151–195.

Junghuhn, F., 1853, Java, deszelfs gedaante, bekleeding en inwendige structuur: Amsterdam, Van Kampen.

—— , 1857, Over de fossiele zoogdierbeenderen te Patihajam, in de residentie Djapara, eiland Java: Natuurkundig Tijdschrift voor Nederlandsch-Indië, v. 14, p. 215–219.

Leinders, J.J.M., Aziz, F., Sondaar, P. Y., and de Vos, J., 1985, The age of the hominid-bearing deposits of Java; State of the art: Geologie en Mijnbouw, v. 64, p. 167–173.

Lydekker, R., 1879, Notices of Siwalik mammals: Records of the Geological Survey of India, v. 12, p. 33–52.

Marks, P., and Drooger, C. W., 1983, Levensbericht van Gustav Heinrich Ralph von Koenigswald (13 November 1902–10 Juli 1982): Amsterdam, Jaarboek Koninklijke Nederlandse Akademie van Wetenschappen, p. 156–163.

Martin, K., 1883, Palaeontologische Ergebnisse von Tiefbohrungen auf Java: Jaarboek Mijnw. Ned.-Indie, 1883, p. 371–412.

—— , 1884, Ueberreste vorweltlicher Proboscidier von Java und Banka: Leiden, Sammlungen des geologischen Reichs-Museums, v. 4, 1884–1889, p. 1–24.

—— , 1887, Fossile Saugethierreste von Java und Japan: Leiden, Sammlungen des geologischen Reichs-Museums, v. 4, 1884–1889, p. 25–69.

—— , 1888, Neue Wirbelthierreste vom Pati-Ajam auf Java: Leiden, Sammlungen des geologischen Reichs-Museums, v. 4, 1884–1889, p. 87–116.

—— , 1890, Ueber neue Stegodon-Reste aus Java: Verhandelingen Koninklijke Nederlandsche Akademie van Wetenschappen, afdeeling Natuurkunde, v. 28, p. 1–13.

Martin, R., 1895, Kritische Bedenken gegen den *Pithecanthropus erectus* Dubois: Globus, v. 67, p. 213–217.

—— , 1896, Weitere Bemerkungen zur *Pithecanthropus*- Frage: Zürich, 18 p.

Martin-Icke, H., 1911, Die fossilen Gastropoden, *in* Selenka, M. L., and Blanckenhorn, M., eds., Die *Pithecanthropus*-Schichten auf Java; Geologische und paläontologische Ergebnisse der Trinil-Expedition (1907 and 1908): Leipzig, Engelmann, p. 46–51.

Matsu'ura, S., 1982, A chronological framing for the Sangiran hominids; Funda-

mental study by the fluorine dating method: Bulletin of the National Science Museum of Tokyo, Series D, Anthropology, v. 8, p. 1–53.
——, 1984, Age of the early Javanese hominids; A review, *in* Else, J. G., and Lee, P. C., eds., Primate evolution, v. 1: Cambridge, Cambridge University Press, p. 115–121.
——, 1985, A consideration of the stratigraphic horizons of hominid finds from Sangiran by the fluorine method, *in* Watanabe, N., and Kadar, D., eds., Quaternary geology of the hominid fossil-bearing formations in Java: Geological Research and Development Centre Special Publication 4, p. 359–366.
Nishimura, S., Ikeda, T., and Yokoyama, T., 1981, A fundamental investigation on the fission track method: Japan, Ministry of Education Report for the Grant-in-Aid Scientific Research, 1980, no. 539009, p. 5–11 (in Japanese).
Oakley, K. P., Campbell, B. G., and Molleson, T. I., 1975, Catalogue of fossil hominids; Part 3, Americas, Asia, Australasia: London, British Museum (Natural History), p. 1–217.
Oppenoorth, W.F.F., 1932, De vondst van palaeolitische menschelijke schedels op Java: De Mijningenieur, v. 6, p. 106–115.
Orchiston, D. W., and Siesser, W. G., 1982, Chronostratigraphy of the Plio-Pleistocene fossil hominids of Java: Modern Quaternary Research Southeast Asia, v. 7, p. 131–149.
Pope, G. G., 1983, Evidence on the age of the Asian Hominidae: Proceedings of the National Academy of Science, v. 80, p. 4988–4992.
Radèn Saléh, 1867, Over . . . fossiele beenderen van den Pandan: Natuurkundig Tijdschrift voor Nederlandsch Indië, v. 29, p. 422–423, 426–429, 433–437, 448–451, 455–459.
Sartono, S., 1979, The age of the vertebrate fossils and artefacts from Cabenge in south Sulawesi, Indonesia: Modern Quaternary Research Southeast Asia, v. 5, p. 68–81.
——, 1982, Characteristics and chronology of early men in Java: Congrès Internationale de Paléontologie Humaine, I, Nice, Pretirage, p. 491–533.
Schlosser, M., 1903, Die fossilen Säugetiere China's nebst einer Odontographie der rezenten Antilopen: Abhandlungen der Königlich Bayerischen Akademie der Wissenschaften, II. Classe, v. 22, I. Abtheilung, 20 p.
Schmulling, P.E.C., 1864, Bij mijne . . . Kaleoso: Natuurkundig Tijdschrift voor Nederlandsch-Indië, v. 27, p. 399–400.
Selenka, M. L., and Blanckenhorn, M., eds., 1911, Die *Pithecanthropus*-Schichten auf Java; Geologische und paläontologische Ergebnisse der Trinil-Expedition (1907 und 1908): Leipzig, Engelmann, 268 p.
Semah, F., 1982, Pliocene and Pleistocene geomagnetic reversals recorded in the Gemolong and Sangiran domes (central Java): Modern Quaternary Research Southeast Asia, v. 7, p. 151–164.
Sloet van Oldruitenborgh, A., 1859, Fossiele beenderen, gevonden op de helling van den Goenung Pandan: Natturkundig Tijdschrift voor Nederlandsch-Indië, v. 16, p. 70–71.
Sondaar, P. Y., 1984, Faunal evolution and the mammalian biostratigraphy of Java: Courier Forschungsinstitut Senckenberg, v. 69, p. 219–235.
Staring, W.C.H., 1865, Het voorkomen van diluviale gronden op Java: Verslagen en Mededeelingen Koninklijke Akademie, v. 17, p. 52.
Stehlin, H. G., 1925, Fossile Säugetiere aus der Gegend von Limbangan (Java): Wetenschappelijke Mededeelingen Dienst Mijnbouw Nederlandsche-Indië, v. 3, p. 1–10.
Stehn, C. E., and Umbgrove, J.H.F., 1929, Bijdrage tot de geologie der vlakte van Bandoeng: Tijdschrift Koninklijk Nederlandsch Aardrijkskundig Genootschap, v. 46, p. 301–314.
Stremme, H., 1911a, Die Säugetiere mit Ausnahme der Proboscidier, *in* Selenka, M. L., and Blanckenhorn, M., eds., Die *Pithecanthropus*- Schichten auf Java; Geologische und paläontologische Ergebnisse der Trinil-Expedition (1907 und 1908): Leipzig, Engelmann, p. 82–160.
——, 1911b, Die Säugetierfauna der *Pithecanthropus*-Schichten: Neues Jahrbuch für Mineralogie, Geologie und Paläontologie, p. 54–60, 83–89.
Suzuki, M., Budisantoso, W., Saefudin, I., and Itihara, M., 1985, Fission track ages of pumice tuff, tuff layers, and javites of hominid fossil-bearing formations in Java: Geological Research and Development Centre Special Publication 4, p. 309–357.
Theunissen, B., 1989, Eugène Dubois and the ape-man from Java; The history of the first "missing link" and its discoverer: Dordrecht, Kluwer Academic Publications, 216 p.
Tobias, P. V., 1984, The life and work of Professor Dr. G.H.R., von Koenigswald, *in* Kramer, W., ed., Auf den Spuren des Pithecanthropus; Leben und Werk von Professor Dr. Gustav Heinrich Ralph von Koenigswald (1902–1982): Frankfurt am Main, Vorträge einer akademischen Gedächtnisfeier, November 1983, p. 25–96.
van Bemmelen, R. W., 1968, Professor G.H.R. von Koenigswald: Payenborchiana, September 1968, p. 23–26.
van den Brink, L. M., 1982, On the mammal fauna of the Wajak Cave Java (Indonesia): Modern Quaternary Research Southeast Asia, v. 7, p. 177–193.
van der Maarel, F. H., 1932, Contributions to the knowledge of the fossil mammalian fauna of Java: Wetenschappelijke Mededeelingen Dienst Mijnbouw Nederlandsch-Indië, v. 15, p. 1–208.
van Es, L.J.V., 1931, The age of *Pithecanthropus:* The Hague, Nijhoff, 142 p.
von Koenigswald, G.H.R., 1931a, Fossielen uit Chineesche apotheken in west-Java: De Mijningenieur, Bandoeng, p. 189–193.
——, 1931b, *Sinanthropus, Pithecanthropus,* en de ouderdom van de Trinil-lagen: De Mijningenieur, Bandoeng, p. 198–202.
——, 1933, Beitrag zur Kenntniss der fossilen Wirbeltiere Javas, Teil 1: Wetenschappelijke Mededeelingen Dienst Mijnbouw Nederlandsch-Indië, v. 23, p. 1–185.
——, 1934, Zur Stratigraphie des Javanischen Pleistocän: De Ingenieur in Nederlandsch Indië, v. 1, section 4, p. 185–200.
——, 1935, Die fossilen Säugetierfaunen Javas: Proceedings Koninklijke Nederlandsche Akademie van Wetenschappen, v. 38, p. 188–198.
——, 1936a, Een nieuwe *Pithecanthropus* ontdekt: Algemeen Indisch Dagblad, 18 March 1936.
——, 1936b, *Pithecanthropus erectus,* antwoord dr Von Koenigswald: Algemeen Handelsblad, 7 May 1936.
——, 1936c, Erste Mitteilung über einen fossilen Hominiden aus dem Altpleistocän Ostjavas: Proceedings Koninklijke Nederlandsche Akademie van Wetenschappen, v. 39, p. 1000–1009.
——, 1937, *Pithecanthropus* received into the human family; With an introductory note by Sir Arthur Keith, F.R.S.: Illustrated London News, 11 December 1937, p. 1039–1041, 1086.
——, 1939, Das Pleistozän Javas: Quartair, v. 2, p. 28–53.
——, 1940, Neue *Pithecanthropus*-Funde 1936–1938; Ein Beitrag zur Kenntnis der Praehominiden: Wetenschappelijke Mededeelingen Dienst Mijnbouw Nederlandsch-Indië, v. 28, p. 1–232.
——, 1956a, Speurtocht in de prehistorie: Amsterdam, Spieghel, 220 p.
——, 1956b, Remarks on the correlation of mammalian faunas of Java and India and the Plio-Pleistocene boundary: Proceedings Koninklijke Nederlandsche Akademie van Wetenschappen, Series B, v. 59, p. 204–210.
——, 1964, Potassium-argon dates and early man: Report 6, International Congress on the Quaternary, Warsaw, 1961, v. 4, p. 325–327.
——, 1968, Das absolute Alter des *Pithecanthropus erectus* Dubois, *in* Kurth, G., ed., Evolution und Hominisation (2nd ed.): Stuttgart, G. Fischer, p. 195–203.
——, 1975, Early man in Java; Catalogue and problems, *in* Tuttle, R. H., ed., Paleoanthropology, morphology, and paleoecology: The Hague, Paris, Mouton, p. 303–309.
Weidenrich, F., 1939, The classification of fossil hominids and their relations to each other, with special reference to *Sinanthropus:* Bulletin of the Geological Society of China, v. 19, p. 64–75.

Manuscript Accepted by the Society June 6, 1989

Printed in U.S.A.

Geological Society of America
Special Paper 242
1990

The Antevs-Bryan years and the legacy for Paleoindian geochronology

C. Vance Haynes, Jr.
Departments of Anthropology and Geosciences, University of Arizona, Tucson, Arizona 85721

INTRODUCTION

In 1888, the founding year of the Geological Society of America, two geologists were born who would have profound influence on late Quaternary geology and the dating of early man in the New World. Kirk Bryan (Fig. 1) was born in Albuquerque, New Mexico, on 22 July 1888, four months before the birth of Ernst Valdemar Antevs (Fig. 2) in Vartofta, Sweden, on 20 November. Whereas both pursued different geological paths and met on only a few occasions, each made good use of the other's concepts in attempting to establish ages for Paleoindian sites in North America. However, they did not always agree. Each found mutually beneficial scientific relationships with archaeologists that lasted for the rest of their lives. As will be pointed out later, they were not the only geologists working with archaeologists in North America. However, they laid down chronologic schemes that were followed and modified by others. Bryan, who died nearly a quarter century before Antevs, left a legacy of students who carried on the Bryan tradition of innovation and independent thinking. Antevs, while having no protégés, had a long-lasting influence on the concept of postglacial climatic changes.

Figure 1. Kirk Bryan at Awatovi archaeological project, Hopi Country, Arizona, 1939.

THE NORTH AMERICAN VARVE CHRONOLOGY

Having studied Scandinavian varve chronologies under Gerard De Geer, Antevs applied his talents to dating the retreat of late Wisconsin glaciation in the northeast. Three gaps in the varve record had to be filled in by estimates based on other criteria, including estimates for the retreat of Niagara Falls (Antevs, 1928). His chronology for deglaciation beginning after culmination 25,000 years ago was accepted by some as the best estimate available (Flint, 1947) and was used later, with some modification, by Bryan and Ray (1940) in dating the Folsom occupation at Lindenmeier. As will be seen further on, it is this chronology, with a few modifications, that Antevs applied to southwestern sites using a geologic-climatic argument for correlation.

CLIMATIC CHANGE, GLACIATION, AND PLUVIAL LAKES

Antevs' work in the west began in 1922 when John C. Merriam, president of the Carnegie Institution, asked him to correlate Bonneville, Lahontan, and Mono Lake chronologies to the North American glacial sequence. Not finding any usefully varved sequences in the lake beds, he resorted to a climatic basis

Haynes, C. V., Jr., 1990, The Antevs-Bryan years and the legacy for Paleoindian geochronology, *in* Laporte, L. F., ed., 1990, Establishment of a geologic framework for paleoanthropology: Boulder, Colorado, Geological Society of America Special Paper 242.

Figure 2. Ernst Antevs at the Naco Clovis site Arizona, 1953. Arizona State Museum photo 3302 by E. W. Haury.

for correlation, reasoning that lake expansion was due to lower summer temperatures and, less importantly, to greater precipitation penecontemporaneously with glacial expansion (Antevs, 1925). The rise of summer temperatures brought on the retreat of glaciers and the evaporation of lakes. On the basis of European pollen evidence, he believed that postglacial climate became warmer than today's climate in both North America and Europe. He reasoned that because climatic change is global, glaciations on both continents must be synchronous. These concepts persisted throughout his career.

ARROYO PROCESSES

Bryan's interest in Quaternary geology apparently began not with glaciation but with alluvial processes. His concern with arroyo cutting, if not already present, was undoubtedly enhanced by association with Herbert E. Gregory, his major professor at Yale, who was conducting geological studies in Navajo country before World War I. Before receiving his doctorate from Yale in 1920, Bryan, as a geological aid in the U.S. Geological Survey, came under the influence of hydrologist Oscar E. Meinzer in the Ground Water Division. Before World War I, Bryan did fieldwork in the Sacramento Valley, California, in the Columbia Plateau, and in Papago country where he again encountered arroyos (Bryan, 1922; Fig. 3). He found archival accounts indicating that most arroyos had cut their steep-walled channels after 1880 (Bryan, 1925). He reasoned, along with others, that the erosion of gullies was caused by a reduction of protective vegetation due not to overgrazing, as Antevs (1936) later maintained, but to a change in climate to drier times. By interviewing old-timers in New Mexico in the 1920s he augmented archival accounts and land surveyors' notes in establishing when and where flood plains were abandoned in the Rio Puerco (Bryan, 1928) and Rio Salado (Bryan, 1927).

In 1924 and 1925, Bryan, then a visiting lecturer at Harvard, worked on the geology of Chaco Canyon for the Chaco Canyon Archaeological Expedition directed by Neil M. Judd of the Smithsonian Institution and sponsored by the National Geographic Society. Thus began an association with archaeologists in the field that lasted a quarter century. In 1935, Frank H. H. Roberts, Jr., on the Chaco project as a Harvard anthropology doctoral candidate, assisted Bryan in the field (Bryan, 1954, Plate 9). Both met the pioneer photographer of the west, William H. Jackson, at Chaco in July or August 1925 (Fig. 4). Bryan was impressed by Jackson's geological observations while a member of the 1877 Hayden survey, particularly his recognition of the buried (post-Bonito) channel (Jackson, 1878). That Bryan had discovered the channel independently (Fig. 5) is revealed by a lantern slide talk he presented 11 March 1925 at the Cosmos Club for the Washington Academy of Sciences (Bryan, 1926). This was at least four months before meeting Jackson at Chaco. Buried channel deposits in alluvial sequences in the Southwest then became his strongest evidence that prehistoric arroyo cutting was due to changes to drier climate and not to the activities of man (such as by overgrazing; Bryan, 1928, 1940, 1942). The latter was no more than the "trigger pull" that started flood-plain entrenchment a few years earlier than it would have started otherwise. Antevs (1936), on the other hand, always maintained that historic arroyo cutting was due strictly to overgrazing and other human-induced causes. However, both were in complete agreement on drought as the cause of erosion of buried channels.

In "Flood-water Farming" Bryan (1929a) suggested prehistoric arroyo cutting and the attendant loss of flood plain as a likely cause of abandonment of the great pueblos.

BRYAN AND PALEOINDIAN GEOCHRONOLOGY

In July 1928, in response to telegrams from American Museum vertebrate paleontologist Barnum Brown (Fig. 6) and Smithsonian archaeologist Neil M. Judd, Alexander Wetmore of the Smithsonian Institution telegraphed Kirk Bryan, then at home in Albuquerque, to proceed to Folsom, New Mexico. This was to examine the important finds near there of fluted projectile points in direct association with skeletons of an extinct species of bison (Wormington, 1957). Bryan stayed overnight in Raton at the same hotel as Neil Judd and Frank Roberts, but not realizing they were there, he left for Folsom the next morning ahead of them (Smithsonian Archives, Record Unit 192, Box 353). In a letter dated 3 August to Wetmore, and having witnessed three Folsom points in place, he reported that the association was valid and said the site is "extremely important and will give a new impetus to a study of the Pleistocene and early phases of the Recent. The emphasis that this will give to the vertebrate paleontology of the Pleistocene will be reflected in much more accurate work in physiography—a gain apart from archaeology" (Smithsonian Archives, Record Unit 192, Box 353).

In reporting on the geology of the type Folsom site, Bryan (1929b, 1937) noted that the arroyo contained two alluvial fills, an older one containing the archaeological associations and calcium carbonate nodules and a younger fill, with an erosional contact between them. He made no attempt either then or later to correlate these deposits to the alluvial chronology, presumably because he felt that the several volcanic flows into the local drainage system interjected a control on sedimentation other than climate.

Figure 3. Kirk Bryan as a "geological aid" of the U.S. Geological Survey in Papago Country, 1917. Kirk Bryan, Jr., states: "He was very proud that he was the first member of the survey to use a Model T Ford for fieldwork in that desert country." Photographer unknown.

Figure 4. Pueblo Bonito, Chaco Canyon, New Mexico, 1925. Individuals from left to right are archaeologists Monroe Amsden and Lynn Hammond, geologist Kirk Bryan, archaeologist Frank H. H. Roberts, Jr., "Picturemaker of the West" William H. Jackson, archaeologist Karl Ruppert, unknown, archaeologist Neil M. Judd, and agriculturalist C. S. Scofield. Photograph probably by O. C. Havens.

ANTEVS AND PALEOINDIAN GEOCHRONOLOGY

Antevs' first experience in dating a Paleoindian site appears to have been in 1934. With support from John Merriam at the Carnegie (Meltzer, personal communication, 1988), Ernst Antevs accompanied Paleoindian specialist E. B. Howard to Blackwater Draw near Clovis, New Mexico, to determine the age of the Clovis occupation associated with diatomaceous pond sediments containing numerous bones of mammoth and extinct bison. He correlated the artifacts and fauna-bearing lake beds at the Clovis site to pluvial Lake Estancia (Meinzer, 1911), 257 km to the west. On the basis of evaporation and runoff calculations, he concluded the lake had reached a high stand approximately 15,000 years ago, after the culmination of Wisconsin glaciation. The artifacts, having been deposited toward the end of pond deposition, he estimated to be between 13,000 and 12,000 years old (Antevs, 1935a).

A GEOLOGIC-CLIMATIC MODEL

In addition to dating Paleoindian sites, Antevs was also interested in the broader aspects of the peopling of the New World. In his paper "The spread of aboriginal man" Antevs (1935b) incorporated W. A. Johnston's (1933) ice-free corridor

Figure 5. Bryan at the post-Bonito channel, Chaco Canyon, New Mexico, 1924. Photograph by O. C. Havens.

hypothesis and dated the corridor opening to between 20,000 and 15,000 years ago. He considered Folsom and Yuma artifacts to be about 12,000 years old and took a conservative view regarding the validity or age of earlier sites. It is here that he began formulating his climatic model for deglaciation or post-pluvial climatic changes in the west. Presumably on the basis of European pollen records (Antevs, 1925), he stated that deglacial temperatures reached the present level (today's) between 9,000 and 8,000 years ago and exceeded today's 7,500 to 4,000 years ago in northern Europe. Because climate is a global phenomenon, he reasoned that North America experienced similar climatic trends. Temperatures, he reasoned, returned to modern levels about 3,500 years ago, staying constant thereafter except for a brief cooling between 500 BC and AD 800. He further suggested that a decrease in precipitation over ice sheets was probably made up for by an increase in deserts. As ice sheets and anticyclones waned, temperatures rose, North Atlantic and North Pacific lows moved northward, and present climate in the Southwest was attained 15,000 to 10,000 years ago. This is essentially the climatic model that, with subsequent modifications and refinement, would eventually be melded with the alluvial chronology being worked out by Bryan and his students.

GEOCHRONOLOGY OF THE COCHISE CULTURE

In 1936, Antevs joined archaeologists Emil Haury and E. B. (Ted) Sayles in examining archaeological sites in alluvium exposed in the banks of Whitewater Draw, Arizona (Fig. 7), where ten years earlier Byron Cummings, then director of the Arizona State Museum, investigated a mammoth skull overlying artifacts at Double Adobe. As a result, Antevs and Sayles developed a lifelong collaboration in defining and dating the Cochise culture (Antevs, 1937; Sayles and Antevs, 1941; Sayles, 1983). In their 1941 opus, Antevs cautiously correlated the alluvial sequence of southeast Arizona to the alluvial sequences that Bryan and his students had worked out in the Davis Mountains area in Texas (Albritton and Bryan, 1939) and Hopi country (Hack, 1939).

GEOCHRONOLOGY OF THE FOLSOM OCCUPATION OF LINDENMEIER

At the same time that Sayles and Antevs were working out the Cochise chronology, Bryan, with students John T. Hack, Franklin T. McCann, and Louis L. Ray, was working with Smithsonian archaeologist Frank H. H. Roberts on the geology of the Lindenmeier Folsom camp site in northern Colorado (Fig. 8). Bryan and Ray (1940) devised an elegant but necessarily complicated way of geologically dating the Folsom occupation. They first worked out the terrace sequence of the Colorado piedmont. However, because the Lindenmeier site lay in an isolated valley, they had to correlate the Folsom level at Lindenmeier to the Clovis level at Dent in order to plug Folsom time into the terrace sequence. No chronological distinction between Folsom and Clovis was recognized at that time. Dent was considered to be a remnant of the 9-m terrace of the South Platte River as mapped at Kersey, 27 km downstream, where Folsom artifacts had been found in dunes on the Kersey terrace. Thus, Folsom occupation of the region was linked to the top of the Kersey terrace. This was correlated to the 8-m terrace of the Cache La Poudre River where Ray (1940) had traced the terraces to glacial moraines in the Rocky Mountains. Thus, the outwash terrace of the Corral Creek moraine was correlated with the Kersey terrace and, therefore, to the Folsom occupation. The Corral Creek was considered to be equivalent to the third substage (WIII) of Wisconsin glaciation.

To obtain an estimate of absolute age they had to incorporate Antevs' (1931) varve chronology. In doing so, they modified some of his estimates for gaps in the varve record and concluded

Figure 6. Barnum Brown and Karl Schwachheim at the Folsom site, New Mexico, 1927. Schwackheim, left, pointing to Folsom point in situ. (Photo courtesy of the American Museum of Natural History.)

Figure 7. Left to right are Winefred Gladwin, Ada Antevs, Edith Sangster (Gila Pueblo secretary), Ernst Antevs, and Harold S. Gladwin (director of Gila Pueblo) at Double Adobe, Arizona, 1935. Mrs. Gladwin's dog, Pattie, in foreground. Photograph by E. B. Sayles. Arizona State Museum photo 75406.

that the Folsom occupation occurred between 10,000 and 25,000 years ago but more likely closer to the latter. Antevs (1941) disagreed with their correlation of Corral Creek and considered it equivalent to the fourth and last substage (WIV) of Wisconsin glaciation. Therefore, he assigned an age of 10,000 years, or a little more, to the Folsom occupation. These complex procedures well exemplify the tenuous linkages necessary to establish the time of human presence before the development of radiocarbon dating (Libby, 1955).

DISAGREEMENT OVER THE GEOCHRONOLOGY OF MINNESOTA "MINNIE"

In June 1937, both Antevs and Bryan, along with Franklin T. McCann and glacial geologist Paul MacClintock, attended a field conference at the site of Minnesota "Man" where, during road construction, the skeleton of a 15-year-old girl had been found approximately 3 m below the surface. If it was in situ in varved lacustrine beds, the find would be contemporary with proglacial Pelican Lake. Earlier, Antevs (1935b), on the basis of photographs, had suggested that the burial was due to a landslide postdating the lake beds. Bryan (1935), after a visit to the site in 1935, concluded that there was no evidence of intrusive burial. The field conference did not resolve the issue, mainly because the road-cutting equipment had stripped off the overlying sediments and no trained observers witnessed the find during removal by the workmen (Bryan and MacClintock, 1938; Antevs, 1938a). Griffin (1965) has stated that the shell pendant found with the skeleton is similar to Late Archaic specimens dated to approximately 4,000 yr BP in Kentucky. A bone sample yielded an unpublished date of 5,000 yr BP (W-530) that Griffin (1965) believed is too young because all of the shellac preservative may not have been removed. It appears unlikely that the skeleton is 12,000 to 13,000 years old as it would be if it were contemporary with the varved sediments of Glacial Lake Pelican.

ANTEVS REFINES HIS GEOLOGIC-CLIMATIC MODEL

In "Postpluvial climatic variations in the Southwest" Antevs (1938b) reviewed the chronologies of Abert and Summer Lakes,

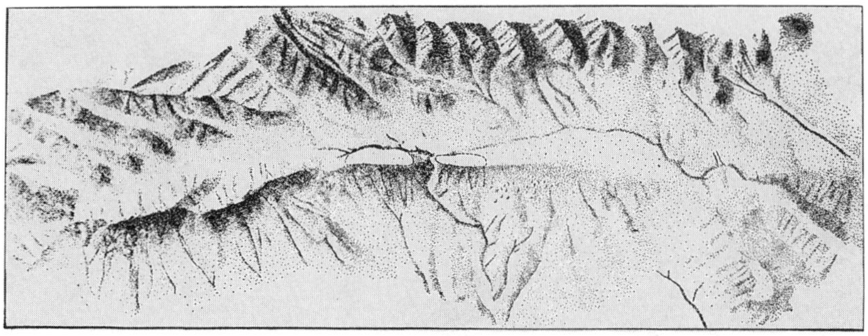

Figure 8. Shaded relief map of the Lindenmeier Valley, Colorado (from Bryan, 1937). The Folsom site is exposed by the gully in the center of the sketch.

Oregon, based on the salinity calculations of Van Winkle (1914), who concluded that the present lake waters had salinities too low to be remnants of the late Pleistocene lake and therefore had to be formed after salts from the evaporation thereof had been deflated during a period warmer and drier than today. He figured the salinities of the two lakes required at least 4,000 years to reach their modern concentrations. Antevs applied this value to date the end of dry conditions and the beginning of modern moist conditions and found confirmation in similar determinations by Gale (1915) for Owens Lake, California. The beginning of this Middle Postpluvial warm dry stage he placed at 7,500 years ago based on pollen chronologies of Swedish bogs (Grandlund, 1932). The beginning of his Early Postpluvial he placed at 10,000 years ago, and the Late Postpluvial covered from 4,000 years ago to the present. The result was an early version of what was to become his model of "Neothermal" climate.

In his "Climatic changes and pre-white man" Antevs (1948) defined the Postpluvial as the Neothermal Age and subdivided it into the Anathermal, cooler and wetter than today, 9,000 to 7,000 years ago; Altithermal, warmer and drier, 7,000 to 4,500 years ago; and Medithermal, modern climate, 4,500 years ago to the present. Considering time estimates for the recession of Niagara Falls to no longer be valid, he reestimated the culmination of Wisconsin glaciation at 20,000 years ago instead of 25,000. The end of Wisconsin glaciation he equated with the beginning of ice retreat from the Salpausselkä II moraine in Finland at 10,150 varve years ago. He believed the retreat from the Cochrane moraine in Ontario, marking the end of WIV, began at the same time because of his conviction that climatic change was global and glaciation therefore was synchronous. This was a major flaw in his telecorrelations and the main cause of his disagreements with radiocarbon dates.

ANTEVS VERSUS RADIOCARBON DATING

By the time of his "Geochronology of the Deglacial and Neothermal Ages" (Antevs, 1953a) radiocarbon dating had been applied to the Two Creeks forest bed. However, he could not accept the possibility that glaciation in Scandinavia ended before it ended in North America. Thus began a long feud with radiocarbon dating as applied to the glacial and pluvial lake chronologies in North America and the alluvial and Paleoindian chronology in the Southwest.

GEOLOGIC-CLIMATIC DATING IN FINAL FORM

In "Geologic-climatic dating in the West" Antevs (1955) brought together all of his concepts of glaciation, deglaciation, climate, alluvial processes, and chronology by long and detailed reanalysis of all previous bases as well as data current at the time. However, he was highly skeptical of all but one radiocarbon date and offered detailed explanations of why he rejected various pertinent radiocarbon dates.

The Anathermal he defined as a time of July temperatures rising to those of today, beginning with glacial retreat 10,150 years ago, the terminal position of Fennoscandian glaciation. The beginning of the Altithermal he shifted back to 7,500 yr BP, on the basis of Swedish pollen analyses (Fries, 1951). The Medithermal remained unchanged from his previous views of late postpluvial temperatures being as today's.

He provided a more detailed model of arroyo processes (Antevs, 1952) in relation to climate and vegetation changes, stating that arroyos are cut when the climate has just turned arid from a previous condition of semiarid. These drought events he correlated with the erosional events of the alluvial chronology of Bryan and his students. The Altithermal was equated with the erosion and dune formation between Depositions 1 and 2 and was called the Long Drought. The erosional interval within Deposition 2 he named the Fairbank Drought and accepted a radiocarbon date of 2,463 ± 310 yr BP (C-519). The erosional break between Depositions 2 and 3 he called the Whitewater Drought, with an estimated age of AD 330. The tree-ring drought of AD 1276 to 1299, the Great Drought of Douglass (1935), he placed within Deposition 3.

BRYAN WORKS WITH PALEOSOLS

While Antevs was formulating his geologic-climatic dating model in the mid-1930s, New Mexico archaeologists Frank C. Hibben and Wesley Bliss were excavating in Sandia Cave, New Mexico. They found what appeared to be a pre-Folsom occupation exemplified by the single shouldered Sandia point. At

Figure 9. Bryan and students at Awatovi archaeological project, Hopi Country, Arizona, 1939. Individuals from left to right are Charles E. Stearns, John T. Hack, Kirk Bryan, unknown, Doak Cox, Al Marston, Hadda Cabot, E. C. (Ted) Cabot, and Franklin T. McCann.

Hibben's invitation, Kirk Bryan studied the geology of Sandia Cave. He interpreted the cave stratigraphy in terms of late Quaternary paleoecological changes by observing compound paleosols that reflected elevations shifts of vegetation zones in the Sandia Mountains (Bryan, 1941a). The limonite ocher stratum, then thought to be separate the Sandia level from the Folsom level, was correlated to a pedalfer on the mountain side that reflected a spruce-fir forest under full glacial climate of 25,000 years ago. Subsequent superimposition of a pedocal upon the pedalfer implied a change of climate to less humid conditions, leading to the pinyon-juniper woodland of today. This was one of the first applications of paleosols to paleoenvironmental reconstruction in North America.

Further observations on polygenetic paleosols were provided by Bryan and Albritton (1943). A complex paleosol in the Davis Mountains of western Texas was interpreted as indicating three stages of dry climate with two intervening moist intervals. They considered that the inferred changes of climate supported their previous interpretations of the relationship between climate and alluvial processes where deposition occurred during relatively humid times and arroyo cutting during relatively arid times (Albritton and Bryan, 1939).

MORE BRYAN AND OTHERS ON ALLUVIAL CHRONOLOGY

In his paper "The geologic antiquity of man in America" Bryan (1941b) discussed all the problems of the peopling of the New World, from crossing the Bering Strait, to coastal migration, to an ice-free corridor through Canada, to shipwrecked Asians or Polynesians. He elaborated on how dating required close integration of archaeology, paleontology, and physical anthropology with geology. He referred to Holocene arroyo cutting and filling in the Southwest as the "Alluvial Chronology." As near as I can determine, this is the first reference to the alluvial chronology in a formal sense.

Near Grants, New Mexico, in 1940, Bryan and McCann (1943) investigated archaeological sites on ancient sand dunes that could be correlated to the alluvial sequence of the area. The placing of the preceramic and post-Paleoindian San Jose complex at the beginning of Deposition 2 filled an important gap in the archaeological record. Also, they became the first of the Bryan school to correlate the alluvial chronology to the Whitewater Draw sequence of Sayles and Antevs (1941). Another Bryan student, Robert L. Nichols (1934), had earlier worked out the volcanic chronology of the Grants area.

The San Jon site in eastern New Mexico was visited by Bryan and Frank H. H. Roberts briefly in 1939 and 1940. The site consisted of a stratified sequence of alluvium and pond sediments exposed in a High Plains depression (Judson, 1950) that had been breeched by the retreat of the High Plains escarpment. Sheldon Judson undertook the geological investigations the final phascs of which had to bc postponcd until aftcr World War II (Judson, 1953). At this time Bryan (1943) and many of his students became involved in military geology.

THE ANTEVS-BRYAN LEGACY

Antevs had no students in the formal sense. While working in archaeological sites he preferred to be left alone until he had worked out the stratigraphy. Only then would he come forth with his interpretations (Agogino, personal communication, 1958). Bryan, on the other hand, frequently had several students about him in the field and was always quite outspoken.

In the 1930s Bryan had a number of graduate students working on Quaternary problems in the West (Fig. 9). The

Figure 10. Archaeologists and geologists relaxing at Awatovi, 1939. From the left to right are Connie Arensberg, John T. Hack, Richard B. Woodbury, Kirk Bryan, Sol Kimball, E. Mott Davis, and Al Lancaster.

names of his students read like a Who's Who in American geomorphology: Claude C. Albritton, Jr.; George H. Crowl; Charles S. Denny; Richard P. Goldthwait; John T. Hack; Sheldon Judson; Franklin T. McCann; Robert L. Nichols; Louis L. Ray; Robert P. Sharp; H. T. U. Smith; Charles E. Stearns; Joseph E. Upson, II; and Herbert E. Wright, Jr. Many worked on archaeological projects, thereby contributing directly to establishing the time of human presence in America. All contributed indirectly to archaeology. During this period, Bryan interacted with Harvard archaeologists (Fig. 10) as well as up-and-coming students of archaeology such as E. Mott Davis, J. Charles Kelley, Richard B. Woodbury, and H. Marie Wormington, to name a few.

Claude C. Albritton (Fig. 11), working with Harvard archaeologist J. Charles Kelley, worked out a three-fold sequence of late Quaternary alluviation in the Trans-Pecos area of Texas and, with Bryan, made the first attempt to correlate alluvial strata over a wide geographic area on the basis of stratigraphy, contained archaeological sites, and remains of extinct fauna (Albritton and Bryan, 1939). This year (1988) the Geological Society of America presented the Archaeological Geology award to Claude Albritton and to Glen L. Evans, as mentioned later.

Harvard's Awatovi Expedition in northern Arizona introduced several Bryan students to a large-scale archaeological project (Fig. 12) and provided subject matter for John Hack's (1942) classic opus on Hopi country. He subdivided Deposition 2 of the original three units into a 2A and 2B on the basis of archaeological content and on an erosional disconformity. Pottery making came in 2A time, whereas 2B is preceramic. (The A and B were later transposed to conform with the order of stratigraphic succession.)

Bryan's work on the geochronology of Ventana Cave with Harvard archaeologist Emil Haury was done in 1942 but not published until after World War II (Bryan, 1950). It was Bryan's last major review of the alluvial chronology and North American glaciation in relation to dating ancient Americans. Bryan (1948) extended the alluvial chronology to the Valley of Mexico in order to estimate the geologic age of vertebrate fossils there.

Sheldon Judson (1953) returned to San Jon in 1947 to finish his field work there and correlated his San Jon sequence to the "Alluvial Chronology."

A new crop of students came under the positive influence of Kirk Bryan after World War II. These included such prominent names in Quaternary geology today as Donald F. Eschman, Joseph H. Hartshorn, Luna B. Leopold, John P. Miller, Harold E. Malde, John H. Moss, Sherwood D. Tuttle, Clyde Wahrhaftig, and M. Gordon Wolman. More up-and-coming archaeologists also took advantage of Bryan's teachings at Harvard, including Herbert W. Dick, Raymond H. Thompson, and Fred Wendorf. H. Marie Wormington (1957) was already well established as a leader in Paleoindian studies when she returned to Harvard after World War II to complete her M.A. and Ph.D. degrees in 1950 and 1954. Kirk Bryan was on her master's committee.

In the Gallup area of New Mexico, Leopold and Snyder (1951) extended the Alluvial Chronology to another region, and Miller, with archaeologist Fred Wendorf (1958), extended it farther east to the Tesuque Valley. Leopold and Miller (1954) teamed up in the late 1940s and early 1950s to produce a monograph on the alluvial chronology of central and eastern Wyoming. They included paleosols as a part of the stratigraphic sequence, and Leopold (1951) brought his training as a meterologist to bear in attributing arroyo cutting to a reduction in frequency of small rains relative to large rains. They correlated their postglacial sequence to the Alluvial Chronology, thus extending it to another region of semiarid climate 800 km to the north.

In 1946 and 1947, John H. Moss (1951) picked up where John T. Hack (1943) had left off before the war and applied Bryan and Ray's Lindenmeier dating technique to dating the Cody Complex bison kill at the Finley site near Eden, Wyoming. Hack had suggested correlating the archaeological horizon in the Kilpecker Dunes to the local terrace sequence and tracing this to glacial moraines in the Wind River Mountains about 80 km to the north. From a reconnaissance, Hack concluded that the Finley site occupation was younger than the Farson terrace and therefore younger than the Temple Lake moraine or Wisconsin IV. This made the Eden points (they were then called "Yuma") considerably younger than Folsom.

Moss (1951) found the Cody occupation to best correlate with the formation of the Paker Terrace, which he traced to the Temple Lake moraine, thereby arriving at the same postglacial correlation as Hack. Thus a post-Folsom age for the Cody complex was established for the first time.

Moss continued his work with Paleoindian archaeological excavations at the Horner site near Cody, Wyoming, from which the name Cody complex was taken. Sheldon Judson also participated, and although neither of their reports were published, they have been used in recent investigations of the Horner site by Frison and Todd (1987).

It was during this work near Cody that Kirk Bryan had a fatal heart attack 21 August 1950. Bryan's legacy continues through his scientific contributions and those of his many students. Antevs died at his home in Globe, Arizona, 19 May 1974, after a prolonged illness (Smiley, 1974). His publications provide a lasting legacy, and his geologic-climatic model remains a basic framework for dating archaeological sites in the American West.

Figure 11. Bryan students Claude C. Albritton, Jr., (left) and J. Fred Smith, Jr., at Sierra Blanca, in the Trans-Pecos area, Texas, 1935.

CONTRIBUTIONS OF CONTEMPORARIES

Antevs and Bryan were not the only ones doing what today we call archaeological geology. From Nebraska to the Southern High Plains other important Paleoindian sites were being studied and placed within geochronological frameworks more or less correlatable with those of Antevs, Bryan, or Bryan's students.

In 1932, vertebrate paleontologist C. Bertrand Schultz (1932) of the Nebraska State Museum reported the occurrence of a Paleoindian projectile point probably associated with the partial skeleton of an extinct form of bison near Cumro in central Nebraska. That same year the now-famous Scottsbluff Bison Quarry was excavated by Schultz and associates (Barbour and Schultz, 1932) and the several projectile points found in direct association with the extinct bison bones were named the Scottsbluff type (Wormington, 1957). Thus began a long and productive relationship of Schultz with the geologists and archaeologists that persists to the present day.

The discovery of the Lime Creek site in 1947 provided Paleoindian strata that could be stratigraphically related to the terrace sequence they had been working on for several years before (Schultz and others, 1951; Schultz and Frankforter, 1948) and became the dissertation subject of Harvard archaeologist E. Mott Davis (1962). Their "Graphic résumé of the Pleistocene of Nebraska" provided a detailed geochronological model for relating vertebrate fossils as well as archaeological finds to the Wisconsin glacial sequence (Schultz and others, 1951). It also provided a framework for one of the first stratigraphic tests of the radiocarbon dating method (Libby, 1955, p. 106–111).

Geologist Elias H. Sellards, former director of the Texas Memorial Museum, devoted a significant part of his life to the study of Paleoindian sites. His 1952 publication "Early Man in America" became an instant classic and was to be complemented by Wormington's 1957 revision of "Ancient Man in North America." Sellards (1938) reported on the association of mammoth bones and Clovis artifacts at the Miami Site in the Texas panhandle. In 1937, geologist Glen L. Evans excavated the site

Figure 12. John T. Hack, Awatovi staff member Carlos, and Kirk Bryan on the "sand road north of the camp," 1939.

located in a small buried High Plains depression. He teamed up with vertebrate paleontologist Grayson E. Meade in 1940 to investigate the Quaternary geochronology of the Texas High Plains (Evans and Mead, 1945). During this pioneering work they discovered the Plainview bison kill site on Running Water Draw in 1944. The locality became the type site for the Plainview projectile point associated with extinct bison (Sellards and others, 1947).

In valleys draining the north and east escarpments of the Llano Estacado (as the Texas High Plains were named during the Spanish Conquest) Evans and Mead correlated two alluvial terraces to Depositions 2 and 3 of Albritton and Bryan (1939) and Hack (1942). However, their interpretation of the relationship of climate to cutting and filling was opposite to that of Antevs (1936) and Bryan (1925).

Beginning in 1949, Sellards and Evans conducted archaeological excavations and geological investigations at the Clovis type site at Blackwater Draw, Locality 1, in eastern New Mexico, a gravel pit that had been abandoned since the excavations of Cotter (1937, 1938). This work resulted in more detailed knowledge of the stratigraphy and the recognition of significant erosional intervals between several units containing the Paleoindian archaeological sites (Evans, 1951). Therefore, it was clearly understood for the first time that the Clovis occupations, and associated mammoth fauna, preceded the Folsom occupation, associated with numerous extinct bison bones but without evidence of mammoth (Sellards, 1952; Sellards and Evans, 1960).

Evans' (1951) discovery of prehistoric wells exposed in the walls of the old gravel pit at the Clovis site provided irrefutable evidence of a major dry episode at the end of Paleoindian occupation. The sequence of (1) desiccation of the lake beds, (2) deflation of a blowout, and (3) prehistoric excavation of wells to a lowered water table, provides convincing evidence that the Altithermal of Antevs was a major dry episode.

Additional contributions to Paleoindian geochronology came as a result of excavations at the Lubbock Reservoir site (Sellards and Evans, 1960) and the Milnesand Site (Sellards, 1955). In 1988 the Geological Society of America presented the Archaeological Geology award to Glen L. Evans who, in accepting the award, acknowledged E. H. Sellards and his influence on archaeological geology.

By the early 1950s, U.S. Geological Society geologists in the Denver area had been taking a closer look at alluvial sequences and had begun to subdivide what had traditionally been mapped as simply Quaternary alluvial (Qal). Charles B. Hunt defined the late Holocene Piney Creek alluvium and correlated it with Deposition 2 of the Alluvial Chronology. Latest Wisconsin gravel and sand he correlated with Deposition 1 (Hunt, 1953, 1954). Harold E. Malde (1955) extended Quaternary mapping to the Louisville Quadrangle north of Denver, and Glen Scott (1960, 1963) made further modifications and correlations from the Kassler Quadrangle southwest of Denver. Scott found a pre–Piney Creek alluvium correlatable to Deposition 2A and correlated Post Piney Creek alluvium with Deposition 3; Piney Creek proper he restricted to Deposition 2B (Scott, 1965). In 1954, Malde (1960) worked with archaeologist Herbert W. Dick in determining the geologic age of the Claypool Site in eastern Colorado. In 1958 he did the same for archaeologist Joe Ben Wheat at the Olsen-Chubbuck site (Malde, 1972). Scott worked with Smithsonian archaeologists Waldo Wedel and Dennis Stanford at Lamb Spring, a Paleoindian site southwest of Denver. At Thorne Cave in Utah, Malde and Schick (1964) described 27 m of alluvial aggradation that occurred between 5,000 and 3,000 yr BP and correlated it to the lower Tsegi (Deposition 2A) of Hack (1942).

In the 1950s, Gerald M. Richmond (1957, 1960) picked up where Louis Ray left off and began a new study of Rocky Mountain glaciation. By making use of paleosols on tills, as well as radiocarbon dating, Richmond developed a new understanding of the glacial chronology (Richmond, 1986).

RADIOCARBON DATING AND THE ALLUVIAL CHRONOLOGY

During the late 1950s and early 1960s, several old sites were revisited to apply radiocarbon dating to archaeological horizons and to better date the alluvial units. In addition, several new sites were located that provided new sequences for correlation. By the mid-1960s, enough radiocarbon dates had been determined from alluvial sequences that a reasonable test of the synchroneity of events could be made (Haynes, 1968). The result was a general degree of time equivalency of the major units of deposition.

Previous investigators had suggested that Deposition 1 in several areas probably contained more than a single depositional event. Radiocarbon dating and geological studies at Clovis sites showed this to indeed be the case. Clovis occupations in Arizona and New Mexico occurred at the end of a long period of high water tables and pluvial lakes correlated with the last pre-Twocreekan glaciation (Mehringer and Haynes, 1965; Haynes, 1975; Haynes and Agogino, 1966). Claims for a pre-Twocreekan age could not be substantiated (Haynes, 1964, 1967). Mammoths did not carry over into Folsom times, which began at the base of a significant depositional unit (B) that was missing at many of the alluvial sites previously studied. The next older alluvial unit (A) is a terrance, commonly higher, representing much larger streams that the channel facies of unit B.

Unit A contains a full Rancholabrean fauna but so far has not produced any artifacts. Unit B was found to date between 11,500 and 7,000 yr BP and to cover the full time span of Paleoindian. More recent dating indicates that the top of unit β (Greek letters are used to avoid confusion of the generalized scheme with specific field designations; Haynes, 1969) is no less than 8,500 years old. Thus, old Deposition 1 is subdivided into units α and β, with the contact being a logical stratigraphic, faunal, and archaeological criterion for placement of the Holocene/Pleistocene boundary at 11,000 yr BP (Haynes, 1984).

The β unit is subdivided into a lower channel facies β_1, and upper fine-grained (silt to fine sand) unit β_2 consisting of various combinations of flood-plain and slope-wash alluvium. The next deposition, unit γ_1, commonly occupies buried arroyo channels and consists of various combinations of channel gravels, slope-wash alluvium, and pond or charco sediments. Being typical of Holocene arroyo fills, it and those that followed, units δ and ϵ, usually cannot be distinguished on lithology alone. Combinations of paleosols, archaeological content, and radiocarbon dating are necessary if the erosional contacts are not apparent. Also, the frequency of cutting and filling increases up sequence such that correlations or testing for equivalent units in different geographic areas requires numerous strategically placed and more precise radiocarbon dates, a situation seldom attained, unfortunately. Furthermore, as the frequency increases, the standard deviations of radiocarbon dates commonly become too large to make convincing tests of contemporaneity. This is especially so considering the fact that variations in the times of arroyo cutting in different geographic areas, as seen in the historic record (Bryan, 1925), shows that at least a century is required for entrenchment to become as general as it is today.

DENOUEMENT

Radiocarbon dating and stratigraphic studies at the Lindenmeier site (Haynes and Agogino, 1960; Haynes and others, 1989) and the Folsom type site (Anderson and Haynes, 1978), as well as at other Folsom sites (Haynes and others, 1984), confidently demonstrates an age range of between 10,000 and 11,000 yr BP for the Folsom culture (Holliday and Johnson, 1986).

Antevs (1941), in equating the Corral Creek stade of Rocky Mountain deglaciation with the Salpauselkä stade (W4) of Scandinavian deglaciation, arrived at an age of about 10,000 varve years for the Folsom occupation at Lindenmeier. This was closer to the radiocarbon age than the estimate of Bryan and Ray (1940) who correlated the Corral Creek with what was then known as the late Mankato stade, now probably equivalent to the Greatlakean (Evenson and others, 1976). If this correlation were valid it would make the Folsom occupation close to 11,000 yr BP. However, our present understanding is that Corral Creek is early Pinedale (W3) or late Wisconsin maximum (Richmond, 1986), which would make Folsom several millenia too old.

Viewing the problem on the basis of radiocarbon dating and archaeology provides a hypothesis for testing. The Kersey terrace has in its fine-grained surficial deposits Paleoindian sites ranging from Clovis (11,000 to 11,500 yr BP) to Cody complex (8,500 to 10,000 yr BP). Its major period of gravel aggradation occurred before 11,000 yr BP and is the result of Late Wisconsin (Pinedale) deglaciation of the Rocky Mountains, which ended before 11,600 yr BP (Richmond, 1986). If this interpretation is correct, then both the Long Draw and Corral Creek moraines correlate with Kersey terrace aggradation, or perhaps a combination of Pleasant Valley and Kersey terrace aggradation, considering Scott's (1963) interpretation and Holliday's (1987) observation that the former merges with the latter downstream from the mouth of the Cache La Poudre. In any case, the Kuner terrace might correlate with the Holocene Satanta Peak moraines of Benedict (1985). In this case the riparian habitat along the South Platte River utilized by Clovis folk would have been at a degraded stage during or following entrenchment of the Kersey terrace but before aggradation of the Kuner terrace. If so, Paleoindian sites from 11,000 to 8,000 yr BP could be expected in stratigraphic succession in Kuner terrace alluvium. This interpretation would be consistent with Paleoindian successions elsewhere (Haynes, 1968, 1984). However, it varies somewhat from the interpretation of Malde (1984) and considerably from that of Holliday (1987) in suggesting a greater degree of South Platte entrenchment at the time of Paleoindian occupation of the Kersey terrace. Critical to solving this problem would be accurate knowledge of the age of the base of the Kuner terrace. Also, the stratigraphic context of Clovis artifacts on the Kersey terrace is important to learn. If Clovis artifacts had occurred in situ at the Frazier site, I would expect them to be on the contact between the "Gley Layer" and the alluvial sand and gravel.

SUMMARY AND CONCLUSIONS

Antevs and Bryan, independent of each other, developed geologic-climatic techniques for dating Paleoindian sites that are generally complementary. Disagreements were based mainly on differences in correlation to glacial chronologies and to the interpretation of the varve record as pioneered by Antevs. Considering what we know today about the geochronology of glaciation, Paleoindian archaeology, and Pleistocene extinction, it is remarkable how well they did without the aid of radiocarbon dating.

Antevs' concept of the Altithermal and the timing thereof was remarkably perceptive and, in spite of some challenges (Antevs, 1962; Martin, 1963), has withstood the test of time. His dating of the Naco Clovis site in Arizona at 11,000 yr BP, in spite of his miscorrelation of Cochrane with Salpausselkä, was right on target because he also correlated the event with the end of the Estancia Pluvial and the Younger Dryas (Antevs, 1953b). A few years later, in dating the Lehner Clovis site, he vacillated between accepting the radiocarbon dates of about 11,300 yr BP or relying on a complex geologic-climatic argument that made the site 13,000 yr BP or older. He finally settled on the latter interpretation (Antevs, 1959).

Whereas the causes of arroyo cutting and filling are still debated, Bryan and his students recognized the existence of a widespread alluvial sequence that remains a useful framework for paleoenvironmental reconstruction as well as for dating the times of human presence. Their use of alpine glacial sequences and fluvial terraces to date Paleoindian sites was, at the time, the most logical way to proceed. In addition to reasonably accurate age estimates for archaeological sites, Quaternary geology has been greatly enhanced by their quests for knowledge.

ACKNOWLEDGMENTS

This research has been supported in part by the Environmental Sciences program of the National Science Foundation and the National Geographic Society. Biographical data on Ernst Antevs were provided by T. L. Smiley (1974) and photographs by Kathy Hubenschmidt, the Arizona State Museum. Biographical data and photographs of Kirk Bryan were provided by Kirk Bryan, Jr., and Margaret Bryan Davis. Additional photographs were provided by John T. Hack and Watson Smith. Correspondence with many of Bryan's former students was of great help in identifying individuals in photographs and in providing personal insights into the supportive and loquacious character of Kirk Bryan. These include Claude C. Albritton, Jr., Charles S. Denny, John T. Hack, Sheldon Judson, Robert L. Nichols, Charles C. Rich, Robert P. Sharp, Charles E. Stearns, Joseph E. Upson II, and Herbert E. Wright, Jr., Harvard plant geographer Hugh M. Raup, long-time friend and associate of Bryan, provided additional insight into Bryan's personality and reaffirmation of his tendency to cross over disciplinary boundaries in the pursuit of knowledge, encouraging students to do likewise.

Of particular help on the personalities and character of both Antevs and Bryan were conversations with Bryant Bannister, Emil Haury, T. L. Smiley, Watson Smith, Raymond H. Thompson, Fred Wendorf, and H. Marie Wormington. Additional help in photo identification was provided by Emil Haury and Frances Joan Mathien. Photocopies of telegrams and correspondence regarding Bryan's 1928 trip to the Folsom site were provided by Susan Libby Glenn, Smithsonian Institution Archives, and David J. Meltzer. Constructive comments to the manuscript were provided by Adrienne Anderson, Julio Betancourt, David J. Meltzer, and Watson Smith. Expert word processing was provided by Doris Sample and expert photo reproduction by James B. Abbott. To all of the above, it is a pleasure to take this opportunity to express my gratitude.

REFERENCES CITED

Albritton, C. C., Jr., and Bryan, K., 1939, Quaternary stratigraphy in the Davis Mountains, Trans-Pecos, Texas: Geological Society of America Bulletin, v. 50, p. 1423–1474.

Anderson, A. B., and Haynes, C. V., Jr., 1978, How old is Capulin Mountain?; Correlation between Capulin Mountain volcanic flows and the Folsom type site, northeastern New Mexico, *in* Linn, R., ed., Proceedings, 1st Conference on Scientific Research in the National Park Service, v. 2: National Park Service Transactions and Proceedings Series 5, U.S. Department of the Interior Government Printing Office, p. 893–899.

Antevs, E., 1925, On the Pleistocene history of the Great Basin: Carnegie Institute of Washington Publication 352, p. 115–153.

——, 1928, The last glaciation: American Geographical Society Research Series 17, 292 p.

——, 1931, Late glacial clay chronology of North America: Smithsonian Report for 1931, p. 313–324.

——, 1935a, The occurrence of flints and extinct animals in pluvial deposits near Clovis, New Mexico; Part 2, Age of Clovis lake beds: Proceedings of the Philadelphia Academy of Natural Science, v. 87, p. 304–311.

——, 1935b, The spread of aboriginal man to North America: Geographical Review, v. 25, p. 302–309.

——, 1936, Dating records of early man in the Southwest: The American Naturalist, v. 70, p. 331–336.

——, 1937, Climate and early man in North America, *in* MacCurdy, G. G., ed., Early man: Philadelphia, Pennsylvania, Lippincott, p. 125–132.

——, 1938a, Was "Minnesota Girl" buried in a gully?: Journal of Geology, v. 46, p. 293–295.

——, 1938b, Postpluvial climatic variations in the southwest: Bulletin of the American Meterological Society, v. 19, p. 190–193.

——, 1941, Age of the Cochise culture stages: Medallion Papers, v. 29, p. 31–56.

——, 1948, The Great Basin, with emphasis on glacial and postglacial times; 3, Climatic changes and pre-white man: University of Utah Bulletin, v. 38, no. 20, p. 168–191.

——, 1952, Arroyo-cutting and filling: Journal of Geology, v. 60, p. 375–385.

——, 1953a, Geochronology of the deglacial and neothermal ages: Journal of Geology, v. 61, p. 195–230.

——, 1953b, Age of the Clovis fluted points with the Naco Mammoth, *in* Artifacts with mammoth remains, Naco, Arizona: American Antiquity, v. 19, p. 15–17.

——, 1955, Geologic-climatic dating in the West: American Antiquity, v. 20, no. 4, part 1, p. 317–335.

——, 1959, Geological age of the Lehner mammoth site: American Antiquity, v. 25, p. 31–34.

——, 1962, Late Quaternary climates in Arizona: American Antiquity, v. 28, p. 193–198.

Barbour, E. H., and Schultz, C. B., 1932, The Scottsbluff bison quarry and its artifacts: Nebraska State Museum Bulletin 34, v. 1, p. 283–286.

Benedict, J. B., 1985, Arapaho Pass: Glacial geology and archaeology at the crest of the Colorado Front Range: Ward, Colorado, Center for Mountain Ar-

chaeology Research Report 3, 182 p.
Bryan, K., 1922, Erosion and sedimentation of the Papago Country, Arizona, with a sketch of the geology, *in* Contributions to the geography of the United States, 1922: U.S. Geological Survey Bulletin 730, p. 19–90.
——, 1925, Date of channel trenching (arroyo cutting) in the arid Southwest: Science, v. 62, p. 338–344.
——, 1926, Recent deposits of Chaco Canyon, New Mexico, in relation to the life of the prehistoric peoples of Pueblo Bonito: Journal of the Washington Academy of Science, v. 16, p. 75–76.
——, 1927, Channel erosion of Rio Salado, Socorro County, New Mexico: U.S. Geological Survey Bulletin 790, p. 17–19.
——, 1928, Historic evidence on changes in the channel of Rio Puerco, a tributary of the Rio Grande in New Mexico: Journal of Geology, v. 36, p. 265–282.
——, 1929a, Flood-water farming: Geographical Review, v. 19, p. 444–456.
——, 1929b, Folsom culture and its age: Geological Society of America Bulletin, v. 40, p. 128–129.
——, 1935, Minnesota man; A discussion of the site: Science, v. 82, p. 170–171.
——, 1937, Geology of the Folsom deposits in New Mexico and Colorado, *in* MacCurely, G. G., ed., Early man: Philadelphia, Pennsylvania, J. B. Lippincott, p. 139–152.
——, 1940, Erosion in the valleys of the Southwest: New Mexico Quarterly, p. 227–232.
——, 1941a, Correlation of the deposits of Sandia Cave, New Mexico, with the glacial chronology, *in* Hibben, F. C., ed., Evidences of early occupation in Sandia Cave, New Mexico, and other early sites in the Sandin-Manzano region: Smithsonian Institution Miscellaneous Collection, v. 99, no. 23, p. 45–64.
——, 1941b, Geologic antiquity of man in America: Science, v. 93, no. 2422, p. 505–514.
——, 1942, Pre-Columbian agriculture in the Southwest as conditioned by periods of alluviation, *in* Native American cultures: Proceedings 8th American Science Congress, p. 57–74.
——, 1943, Geologists in war service: Science, v. 98, p. 300.
——, 1948, Los suelos complejos y fosiles de la altiplanicie de Mexico en relacion a los cambios climaticos: Sociedad Geologica Mexicana Boletin, tomo 13, part 1, p. 1–20.
——, 1950, The geology and fossil vertebrates of Ventana Cave, *in* Haury, E. W., The stratigraphy and archaeology of Ventana Cave: Tucson, University of Arizona Press, p. 75–126.
——, 1954, The geology of Chaco Canyon, New Mexico: Smithsonian Miscellaneous Collection, v. 122, no. 7, p. 1–65.
Bryan, K., and Albritton, C. C., Jr., 1943, Soil phenomena as evidence of climate changes: American Journal of Science, v. 241, p. 469–490.
Bryan, K., and MacClintock, P., 1938, What is implied by "disturbance" at the site of Minnesota man?: Journal of Geology, v. 46, p. 279–292.
Bryan, K., and McCann, F. T., 1943, Sand dunes and alluvium near Grants, New Mexico: American Antiquity, v. 8, no. 3, p. 281–290.
Bryan, K., and Ray, L. L., 1940, Geologic antiquity of the Lindenmeier site in Colorado: Smithsonian Miscellaneous Collections, v. 99, no. 2, 76 p.
Cotter, J. L., 1937, The occurrence of flints and extinct animals in pluvial deposits near Clovis, New Mexico; Part 4, Report on excavations at the gravel pit, 1936: Philadelphia, Pennsylvania, Proceedings of the Philadelphia Academy of Natural Science, v. 89, 16 p.
——, 1938, The occurrence of flints and extinct animals in pluvial deposits near Clovis, New Mexico; Part 6, Report on field season of 1937: Philadelphia, Pennsylvania, Proceedings of the Philadelphia Academy of Natural Science, v. 90, p. 113–117.
Davis, E. M., 1962, Archaeology of the Lime Creek site in southwestern Nebraska: Omaha, University of Nebraska State Museum Special Publication 3, 106 p.
Douglas, A. E., 1935, Dating Pueblo Bonito and other ruins of the Southwest: National Geographic Society Pueblo Bonito Series, no. 1, p. 1–74.
Evans, G. L., 1951, Prehistoric wells in eastern New Mexico: American Antiquity, v. 17, p. 1–8.
Evans, G. L., and Meade, G. E., 1945, Quaternary of the Texas High Plains: Austin, University of Texas Bulletin 4401, p. 485–507.
Evenson, E. B., Farrand, W. R., Mickelson, D. M., Eschman, D. F., and Maher, L. J., 1976, Greatlakean Substage; A replacement for Valderan in the Lake Michigan Basin: Quaternary Research, v. 6, p. 41–424.
Flint, R. F., 1947, Glacial geology and the Pleistocene epoch: New York, John Wiley and Sons, 589 p.
Fries, M., 1951, Pollenanalytische Zengnisse de spatquartaren Vegetationsentwicklung, hanptsachlich der Waldgeschichte, im nordwestlichen Gotaland (Sudschweden): Uppsala, Acata Phytogeographica Suecica, no. 29, 220 p.
Frison, G. C., and Todd, L. C., eds., 1987, The Horner Site; The type site of the Cody cultural complex: Orlando, Florida, Academic Press, 586 p.
Gale, H. S., 1915, Saline in the Owens, Searles, and Panamint basins, southeastern California: U.S. Geological Survey Bulletin 580, p. 251–323.
Grandlund, E., 1932, Die Geologie der schwedischen Hochmoore: Stockholm, Sveriges Geologiska undersökning, Series C, no. 373, Ärsbok 26, p. 169.
Griffin, J. B., 1965, Late Quaternary prehistory in the northeastern Woodlands, *in* Wright, H. E., Jr., and Frey, D. G., eds., The Quaternary of the United States: Princeton, New Jersey, Princeton University Press, p. 661–662.
Hack, J. T., 1939, The late Quaternary history of several valleys of northern Arizona; A preliminary announcement: Museum of Northern Arizona Notes, v. 11, p. 67–73.
——, 1942, The changing physical environment of the Hopi Indians of Arizona: Cambridge, Massachusetts, Harvard University Peabody Museum Papers, v. 35, no. 1, 85 p.
——, 1943, Antiquity of the Finley Site: American Antiquity, v. 8, p. 235–241.
Haynes, C. V., Jr., 1964, Fluted projectile points; Their age and dispersion: Science, v. 145, p. 1408–1413.
——, 1967, Carbon-14 dates and early man in the New World, *in* Martin, P. S., and Wright, H. E., eds., Pleistocene extinctions; The search for a cause; Proceedings, 7th Congress, International Association for Quaternary Research, v. 6: New Haven, Connecticut, Yale University Press, p. 267–286.
——, 1968, Geochronology of late Quaternary alluvium, *in* Morrison, R. B., and Wright, H. E., eds., Means of correlation of Quaternary successions; Proceedings, 7th INQUA Congress: Salt Lake City, University of Utah Press, v. 8, p. 591–631.
——, 1969, Late Quaternary stratigraphy of the southwestern United States and the Pleistocene-Holocene boundary: Paper presented at 8th INQUA Congress; Abstract in Resumes des Communications, p. 208.
——, 1975, Pleistocene and Recent stratigraphy, *in* Wendorf, F., and Hester, J. J., eds., Late Pleistocene environments of the southern High Plains: Fort Burgwin Research Center Publication 9, p. 57–96.
——, 1984, Stratigraphy and late Pleistocene extinction in the United States, *in* Martin, P. S., and Klein, R. G., eds., Quaternary extinctions; A prehistoric revolution: Tucson, University of Arizona Press, 892 p.
Haynes, C. V., Jr., and Agogino, G. A., 1960, Geological significance of a new radiocarbon date from the Lindenmeier site, Colorado: Denver Museum of Natural History Proceedings, v. 9, 23 p.
——, 1966, Prehistoric springs and geochronology of the Clovis site, New Mexico: American Antiquity, v. 31, p. 812–821.
Haynes, C. V., Jr., Donahue, D. J., Jull, A.J.T., and Zabel, T. H., 1984, Application of accelerator dating to fluted point Paleoindian sites: Archaeology of Eastern North America, v. 12, p. 184–191.
Haynes, C. V., Jr., Beukens, R. P., Jull, A.T.J., and Davis, O. K., 1989, New radiocarbon dates for some old Folsom sites using accelerator technology, *in* Stanford, D., and Day, J. S., eds., Ice age hunters of the Rockies: Denver Museum of Natural History publication (in press).
Holliday, V. T., 1987, Geoarchaeology and late Quaternary geomorphology of the middle South Platte River, northeastern Colorado: Geoarchaeology, v. 2, p. 317–329.
Holliday, V. T., and Johnson, E., 1986, Re-evaluation of the first radiocarbon age

for the Folsom culture: American Antiquity, v. 51, p. 332–338.

Hunt, C. B., 1953, Pleistocene-Recent boundary in the Rocky Mountain region: U.S. Geological Survey Bulletin 996–A, 25 p.

—— , 1954, Pleistocene and Recent deposition in the Denver area, Colorado: U.S. Geological Survey Bulletin 995–C, p. 112.

Jackson, W. H., 1878, Report on the ancient ruins examined in 1875 and 1877: U.S. Geological and Geographical Surveys of the Territories for the year 1876, 10th Annual Report, p. 431–450.

Johnston, W. A., 1933, Quaternary geology of North America in relation to the migration of man, *in* Jenness, D., ed., The American aborigines: Toronto, Ontario, University of Toronto Press, p. 11–45.

Judson, S., 1950, Depressions of the northern portion of the Southern High Plains of eastern New Mexico: Geological Society of America Bulletin, v. 61, p. 253–273.

—— , 1953, Geology of the San Jon Site, eastern New Mexico: Smithsonian Miscellaneous Collections, v. 121, no. 1, p. 1–70.

Leopold, L. B., 1951, Rainfall frequency; An aspect of climatic variation: EOS Transactions of the American Geophysical Union v. 32, no. 3, p. 347–357.

Leopold, L. B., and Miller, J. P., 1954, A postglacial chronology for some alluvial valleys in Wyoming: U.S. Geological Survey Water Supply Paper 1261, 90 p.

Leopold, L. B., and Snyder, C. T., 1951, Alluvial fills near Gallup, New Mexico: U.S. Geological Survey Water-Supply Paper 1110–A, 19 p.

Libby, W. F., 1955, Radiocarbon dating: Chicago, University of Chicago Press, 175 p.

Malde, H. E., 195, Surfical geology of the Louisville Quadrangle, Colorado: U.S. Geological Survey Bulletin 996–E, p. 217–259.

—— , 1960, Geological age of the Claypool site, northeastern Colorado: American Antiquity, v. 26, p. 236–243.

—— , 1972, Geology of the Olsen-Chubbuck site, Appendix 1, *in* Wheat, J. B., The Olsen-Chubbuck site; A Paleo-Indian bison kill: Society of American Archaeology Memoir 26, p. 171–177.

—— , 1984, Geology of the Frazier site, Kersey, Colorado, *in* Anderson, A. B., ed., Paleo-indian sites of the Colorado Piedmont: American Quaternary Association Field Trip Guidebook, p. 13–16.

Malde, H. E., and Schick, A. P., 1964, Thron cave, northeastern Utah; Geology: American Antiquity, v. 30, no. 1, p. 60–73.

Martin, P. S., 1963, The last 10,000 years; A fossil pollen record of the American Southwest: Tucson, University of Arizona Press, 87 p.

Mehringer, P. J., Jr., and Haynes, C. V., Jr., 1965, The pollen evidence for the environment of early man and extinct mammals at the Lehner mammoth site, southeastern Arizona: American Antiquity, v. 31, p. 17–23.

Meinzer, O. E., 1911, Geology and water resources of Estancio Valley, New Mexico: U.S. Geological Survey Water-Supply Paper 275, 89 p.

Miller, J. P., and Wendorf, F., 1958, The alluvial chronology of the Tesuque Valley, New Mexico: Journal of Geology, v. 66, p. 177–194.

Moss, J. H., 1951, Early man and the Eden Valley: Philadelphia, University of Pennsylvania Museum Monographs, 124 p.

Nichols, R. L., 1934, Quaternary geology of the San Jose Valley, New Mexico: Geological Society of America Proceedings for 1933, p. 453.

Ray, L. L., 1940, Glacial chronology of the southern Rocky Mountains: Geological Society of America Bulletin, v. 51, p. 1851–1918.

Richmond, G. M., 1957, Three pre-Wisconsin glacial stages in the Rocky Mountains region: Geological Society of America Bulletin, v. 68, p. 239–262.

—— , 1960, Glaciation of the east slope of the Rocky Mountain National Park, Colorado: Geological Society of America Bulletin, v. 71, p. 1371–1382.

—— , 1986, Stratigraphy and correlation of glacial deposits of the Rocky Mountains, the Colorado Plateau, and the Ranges of the Great Basin, *in* Quaternary glaciations in the Northern Hemisphere: Quaternary Science Reviews, v. 5, p. 99–127.

Sayles, E. B., 1983, The Cochise cultural sequence in southeastern Arizona: Tucson, University of Arizona Press Anthropological Papers of the University of Arizona, no. 42, 179 p.

Sayles, E. B., and Antevs, E., 1941, The Cochise culture: Globe, Arizona, Gila Pueblo Medallion Papers 29, 81 p.

Schultz, C. B., 1932, Association of artifacts and extinct mammals in Nebraska: Nebraska Museum Bulletin 33, v. 1, p. 171–183.

Schultz, C. B., and Frankforter, W. D., 1948, Preliminary report on the Lime Creek sites; New evidence of early man in southwestern Nebraska: Lincoln, University of Nebraska State Museum Bulletin, v. 3, p. 43–62.

Schultz, C. B., Lueninghoener, G. C., and Frankforter, W. D., 1951, A graphic résumé of the Pleistocene of Nebraska: Lincoln, University of Nebraska State Museum Bulletin, 41 p.

Scott, G. R., 1960, Subdivision of the Quaternary alluvium east of the Front Range near Denver, Colorado: Geological Society of America Bulletin, v. 71, p. 1541–1543.

—— , 1963, Quaternary geology and geomorphic history of the Kassler Quadrangle, Colorado: U.S. Geological Survey Professional Paper 431–A, 70 p.

—— , 1965, Nonglacial Quaternary geology of the southern and middle Rocky Mountains, *in* Wright, H. E., Jr., and Frey, D. G., eds., The Quaternary of the United States: Princeton, New Jersey, Princeton University Press, p. 243–254.

Sellards, E. H., 1938, Artifacts associated with fossil elephant: Geological Society of America Bulletin, v. 49, p. 999–1010.

—— , 1952, Early man in America: Austin, University of Texas Press, 211 p.

—— , 1955, Fossil bison and associated artifacts from Milnesand, New Mexico: American Antiquity, v. 20, p. 336–344.

Sellards, E. H., and Evans, G. L., 1960, The Paleo-Indian culture succession in the central High Plains of Texas and New Mexico, *in* Men and cultures; Selected papers of the 5th International Congress of Anthropological and Ethnological Sciences, Philadelphia, 1956: Philadelphia, University of Pennsylvania Press, p. 639–647.

Sellards, E. H., Evans, G. L., Meade, G. E., and Krieger, A. D., 1947, Fossil bison and associated artifacts from Plainview, Texas: Geological Society of America Bulletin, v. 58, p. 927–954.

Smiley, T. L., 1974, Memorial to Ernst Valdemar Antevs, 1888–1974: Geological Society of America Memorials 1974, p. 1–7.

Van Winkle, W., 1914, Quality of the surface waters of Oregon: U.S. Geological Survey Water-Supply Paper 263, p. 117–123.

Wormington, H. M., 1957, Ancient man in North America: Denver Museum of Natural History Popular Series, no. 4, 4th ed., 322 p.

Manuscript Accepted by the Society June 6, 1989

Geological Society of America
Special Paper 242
1990

History of the dating of Homo erectus *at Zhoukoudian*

Zhou Ming Zhen
Institute of Vertebrate Paleontology and Palaeoanthropology, Academia Sinica, The People's Republic of China
Ho Chuan Kun
1140 Los Padres Boulevard, Santa Clara, California 95050

INTRODUCTION

The science of vertebrate paleontology in China began with the publication of the great English anatomist Sir Richard Owen's paper entitled "On fossil remains of mammals found in China" in 1870 (Zhou, 1981). From then until 1927, most of the fossils were either purchased from Chinese drugstores or were merely surface material collected by amateurs, and thus localities and horizons of many of these fossils were unknown. In the early 1920s, fossil collections with accurate stratigraphic data were assembled for the first time. The Zhoukoudian hominid site (Locality 1; Fig. 1) was one of those excavated paleontologically in the 1920s and then was dug archaeologically in 1931 due to the important discovery of quartz flakes, burnt bones, fire-cracked rocks, and ash layers in direct association with *Homo erectus* skeletal remains.

With Henry Fairfield Osborn's central Asia theory of human origins in mind, Davidson Black, a former student of the great English comparative anatomist Grafton Elliot Smith, was convinced that Asia was the "cradle of mankind." The discovery of Zhoukoudian undeniably provided Black with solid evidence supporting the hypothesis of the central Asiatic origin of the Hominidae (Black, 1926). Twenty-five years before Black's identification of "*Sinanthropus pekinensis*," a human tooth with red sedimentary matrix still attached to the roots had already been discovered in a Peking drugstore. Because the tooth was found associated with *Hipparion* fossil horse teeth, as well as with other Pliocene fossils, the anthropologist Max Schlosser suggested that the human tooth might have been of Pliocene age; he even predicted that a new fossil anthropoid or Tertiary man might be found (Von Koenigswald, 1983).

This paper traces the history of the discovery and the dating of the Zhoukoudian cave site and discusses the impact of the new temporal framework on the understanding of middle Pleistocene *Homo erectus* hominids.

RELATIVE DATING OF ZHOUKOUDIAN

The geologic framework of the Tertiary and Quaternary periods of Zhoukoudian was first described by Yih in 1920. Shortly after, the Locality 1 cave deposit of Zhoukoudian was discovered and surveyed by Andersson in 1921 and was described by Zdansky in 1923. Based on the presence of a true horse (*Equus*) and the absence of *Hipparion,* Andersson (1923) suggested that the fauna represented at the locality could be late Pliocene in age. The upper strata (units 1 to 8) of Locality 1 were described by the Chinese geologist Li and the Swedish paleontologist Bohlin in 1927. The next year Young and Pei were formally in charge of the excavation. Teilhard de Chardin and Young (1929) subdivided the 35 m of cave deposits into ten lithostratigraphic units of sediments and then included their paleontological assemblages. They suggested that "Choukoutien [Zhoukoudian] beds are distinctly connected with the Sanmenian period and belong surely to the lower Pleistocene. Though on the other hand, they are distinctly younger than the Nihowan [Nihewan] Sanmenian beds along Sankan Ho [Sangganhe]" (1933, p. 61). On the basis of this new division, Black and others published their joint research in a book entitled *Fossil Man in China* (1933). The age of the Zhoukoudian hominid fossils was estimated to be older than the late Pleistocene and younger than the late Pliocene. The deposits that contained assemblages of just artifacts were dated broadly to the middle Pleistocene in terms of European chronologies. In terms of geologic conditions of the site, Zhoukoudian *Homo erectus* was considered by Weidenreich (1936) as living at the beginning of the Pleistocene. In addition, Weidenrich stated the morphological features showed that "they should be placed much nearer the line of evolution of recent man than to that of anthropoids, especially Tertiary deposits in Asiatic soil" (1936, p. 36–37). The hominid status of Zhoukoudian *Homo erectus* was further confirmed after detailed morphological comparisons with Neanderthal man and other anthropoids.

Based on newly discovered strata and faunal assemblages, three more units below the tenth unit were proposed by Jia (1950) in 1949. Thus, the Zhoukoudian Formation expanded from one stratigraphic unit to thirteen, and the cave deposits were regarded as middle Pleistocene instead of lower Pleistocene. Underlying unit 13, a reddish loam gravel was discovered and named the "basal gravel" when the locality was excavated under

Zhou, M. Z., and Ho, C. K., 1990, History of the dating of *Homo erectus* at Zhoukoudian, *in* Laporte, L. F., ed., 1990, Establishment of a geologic framework for paleoanthropology: Boulder, Colorado, Geological Society of America Special Paper 242.

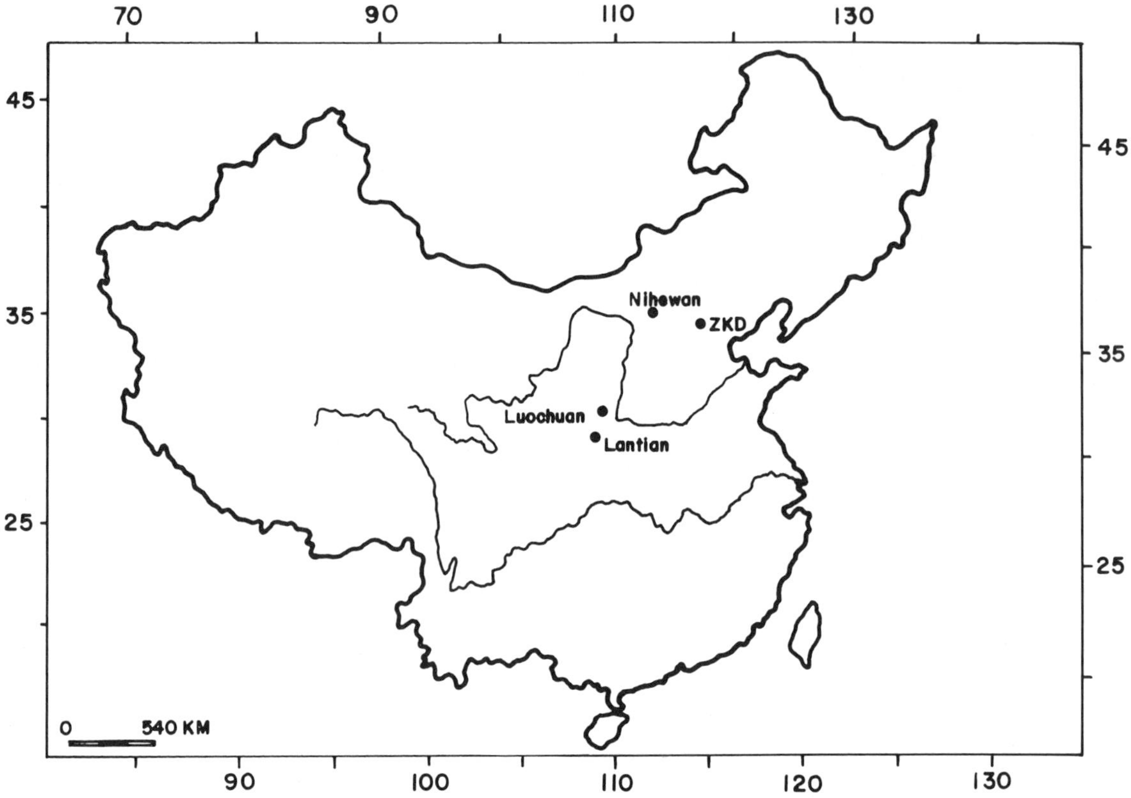

Figure 1. Location map of sites mentioned in the text. ZKD, Zhoukoudian (Choukoutien) site, 40 km southwest of Beijing.

the direction of Jia in 1949 to 1951. Huang (1960) proposed that the cave deposits should be divided solely in terms of lithology in a reappraisal of the site celebrating the thirtieth anniversary of the discovery of the Peking man skull in 1959.

The paleoclimatic setting of Zhoukoudian had been inferred by various workers in terms of faunal composition and sedimentology. Pei (1939) once again pointed out that the date of locality 1 could be broadly contemporaneous with the Cromerian of Western Europe. A sample of pollen extracted from fossil deer specimens (*Sinomegacerus pachyosteus*) preserved in the Uppsala collection was studied, leading Kurten (1960) to suggest that the deposit at Locality 1 could be correlated with the European Elster II (Mindel II), which is approximately 360,000 years old. In 1957, Kahlke and Hu suggested that the cave deposits of Locality 1 correlated with the late second glacial or Holsteinian of European glacial and interglacial chronology. Zhou (1955) inferred that the paleoclimate of Locality 1 could be compared with the warm phase of an interglacial in terms of the vertebrate faunal elements. In 1961, Kahlke and Chow (Zhou) once again argued that the middle and lower strata reflected the transitional phase from a late glacial to interglacial, and that depositional units above the twelfth unit were contemporaneous with the Holsteinian interglacial phase. Jia (1978) indicated that two climatic cycles (from cold to warm) were distinguished in Locality 1 on the basis of the vertebrate assemblages. Zhou (1979) suggested that units 13 to 12 represented deposition during glacial times solely on the basis of rhinocerotid fossils. In 1979, he interpreted units 11 to 8 to represent a transitional phase of an interglacial; units 7 to 5 as typical interglacial; and units 4 to 1 as a cold glacial phase.

These climatic and biostratigraphic correlations have been tested by isotopic dating methods. Employing the uranium and thorium method, Locality 1 has been dated at older than 500 Ka, 240 Ka, and 210 Ka in 1961 by a Russian geochemist, V. V. Cherdintsev, but unfortunately, no stratigraphic information was provided for the samples analyzed (Yang and others, 1985). Palynological and macrofossil plant analyses of the samples from Locality 1 suggested that paleoclimate of Locality 1 was cooler than present (Chaney and Daugherty, 1933; Chaney, 1935; Kurten, 1960). Detailed pollen analysis of samples from units 3 to 11 by Hsu (1965) and Sun (1966) led them to suggest that Zhoukoudian *Homo erectus* was living in an interglacial environment, and also that four climatic oscillations could be delineated. Hsu further indicated that pollen from the basal gravel may reflect a glacial climatic regime. Restudy of samples from the same units resulted in the same conclusion (Kong and others, 1985). In general, most authorities agree that units 3 to 10 were deposited during an interglacial period, but the question of whether the basal gravel was deposited during a glacial or an interglacial climatic regime has not been satisfactorily resolved.

In terms of Oakley's hierarchy of relative dating (1966), the relative age of Zhoukoudian had been estimated either by the age relations between fossil specimens and their associated sediments

(R.1); by the local or regional stratigraphic sequence to which the fossil-bearing horizon can be inferred to belong (R.2); by world stratigraphy (R.3); or even by cranial morphology (R.4). In 1977, the locality was dated by stable isotopes, radiogenic isotopes, and biostratigraphic correlation. The multidisciplinary study of the Zhoukoudian fossil localities, beginning in 1977, undoubtedly can be regarded as the turning point in the application of absolute dating methods in Chinese paleoanthropology (Wu and Wang, 1985).

ABSOLUTE DATING METHODS

Ever since the discovery of the first skull in 1929, most of the Peking fossil homonids were dated either by relative biostratigraphic or lithostratigraphic correlations. Since 1979, six institutes of the Chinese Academy of Science and Chinese Academy of Geological Science, along with 12 laboratories, participated in a multidisciplinary study of the Locality 1 cave. Six dating methods have been used to arrive at a new chronological assessment of this locality. They are carbon 14, uranium series, fission track, thermoluminescence, amino-acid racemization, and plaeomagnetism. Samples were collected systematically from stratigraphic units 1 to 2, 3, 4, 6, 7, 8 to 9, 10 to 11, and 17. Starting from the bottom, the results are briefly described as follows. Unit 11 has been dated to 520 ± 61 Ka by thermoluminescence (Liu and others, 1985); aspidelite in the ash layer of occupation level 10 was dated at 462 ± 15 Ka by fission track (Liu and others, 1985) and older than 417 and younger than 592 Ka by thermoluminescence (Pei and Sun, 1979; Pei, 1985). As part of a scientific exchange program between the People's Republic of China and the United States, enamel samples of fossil teeth were collected for amino-acid racemization analyses from units 8 and 10 to 11 in 1979 (Bada, 1987). The results show that unit 8 was estimated to be 420 Ka, and units 10 to 11 at 500 Ka; by using the uranium-series dating method, units 8 to 9 were also dated to 420 +18/–10 Ka (Zhao and others, 1985). Unit 7 was dated to 370 ± 40 Ka by paleomagnetism (Qian and others, 1979). The upper unit 4 was dated to 256 +62/–40 Ka and 230 +30/–23 Ka by uranium series (Zhao and others, 1985; Chen and Yuan, 1988). Samples of bones and antlers collected from the occupation levels of 1 to 3 have been dated to 292 ± 26 and 312 ± 28 Ka by thermoluminescence (Pei, 1985). Aigner (1986, 1987) was extremely suspicious about the reliability of these dating methods (thermoluminescence, uranium series, and thorium-uranium), but she apparently overlooked the fact that the dates are extremely consistent (Pope, 1988). We thus suggest that the cave of Locality 1 at Zhoukoudian could have been occupied intermittently for some half a million years (Fig. 2).

Paleomagnetic studies show that units below unit 14 (730 Ka) were deposited in the Brunhes Normal Epoch (Liu and others, 1977) and those still farther below were formed in the Matuyama Reversed Epoch (Liu, 1985). It is also interesting to note that unit 7 shows a reversed event at 3.8 Ka, corresponding to Lake Biwa E Event (ca. 400 Ka) within the Brunhes Normal Epoch (Derbyshire, 1983; Liu, 1985; Qian and others, 1985). Recently, Pope (1988) has questioned the presence of such an event but he did not specify the evidence. We suggest that the cave infilling occurred during approximately 730 to 230 Ka. The earlier date is roughly contemporaneous with the upper deposit of Xiehu Formation of Lantian, Shaanxi. Magnetostratigraphic studies show that Lantian *Homo erectus* is much earlier than Zhoukoudian *Homo erectus*. The latest date of Zhoukoudian is close to the bottom level of paleosol no. V of Luochuan Loess (220 Ka dated by thermoluminescence) in Shaanxi, which is the top of lower Lishi Loess (An and Ho, 1989).

Since the results of various different dating methods for Zhoukoudian Locality 1 are rather consistent, Xu and You (1982) and Xu and Ou (1982) attempted to correlate the stratigraphic units from Zhoukoudian with ^{18}O stages. In their opinion, unit 10 (cold) is equivalent to stage 12 (425 Ka); units 8 to 9 to stage 11 (warm); units 6 to 7 to stage 10 (cold); unit 5 to stage 9 (warm); and units 3 to 4 to stage 8 (240 Ka). Recently, Liu (1985, 1988) and Liu and Ding (1984) have also attempted not only to correlate with the ^{18}O curve of Equatorial Pacific core V 28-238 but also with loess stratigraphy. The results show that all units representing the thick ash layers and the travertine layers can be correlated with interglacials, and the units with breccia layers with glacials. Furthermore, those stratigraphic units above unit 14 can also be correlated with deep-sea cores (stages 16 to 6; see Fig. 2).

DISCUSSION

In retrospect, before the discovery of hominid fossils at Zhoukoudian, most hypotheses of homonid origins in Asia were based on geologic settings without reference to *any* hominid fossil record. The ambitious expeditions to the Gobi Desert in search of the first humans ended up with dinosaur eggs; the hypotheses and the expeditions still had great impact with respect to "the very ready acceptance of the most meager fossil evidence—a single tooth—from Choukoutien" (Lewin, 1987). After the discovery of *Homo erectus* at Zhoukoudian, more attention was focused on paleoenvironmental settings and on hominid lifestyles.

Based on Black (1926) and Weidenreich's (1936) description of Zhoukoudian *Homo erectus,* Teilhard de Chardin assumed that such morphological stability as noted by Black and Weidenreich would be evidence for the "slowness which characterizes biological evolution whenever not obscured, disturbed, or accelerated by intrusive immigration of foreign elements" (1936, p. 60). Therefore, early Pleistocene eastern Asia seems to have represented "a quiet and conservative corner amidst the fast advancing human world" (1936, p. 60). Such a view is still held by most Chinese paleoanthropologists (Wu and Lin, 1984). The same phenomenon of "stability" has also been observed by Movius (1948) and Pei and Zhang (1985) among assemblages of artifacts from different stratigraphic units from Zhoukoudian. The question of such assumed stasis in the evolution of *Homo erectus* has, however, been raised recently by Wolpoff (1984) by using

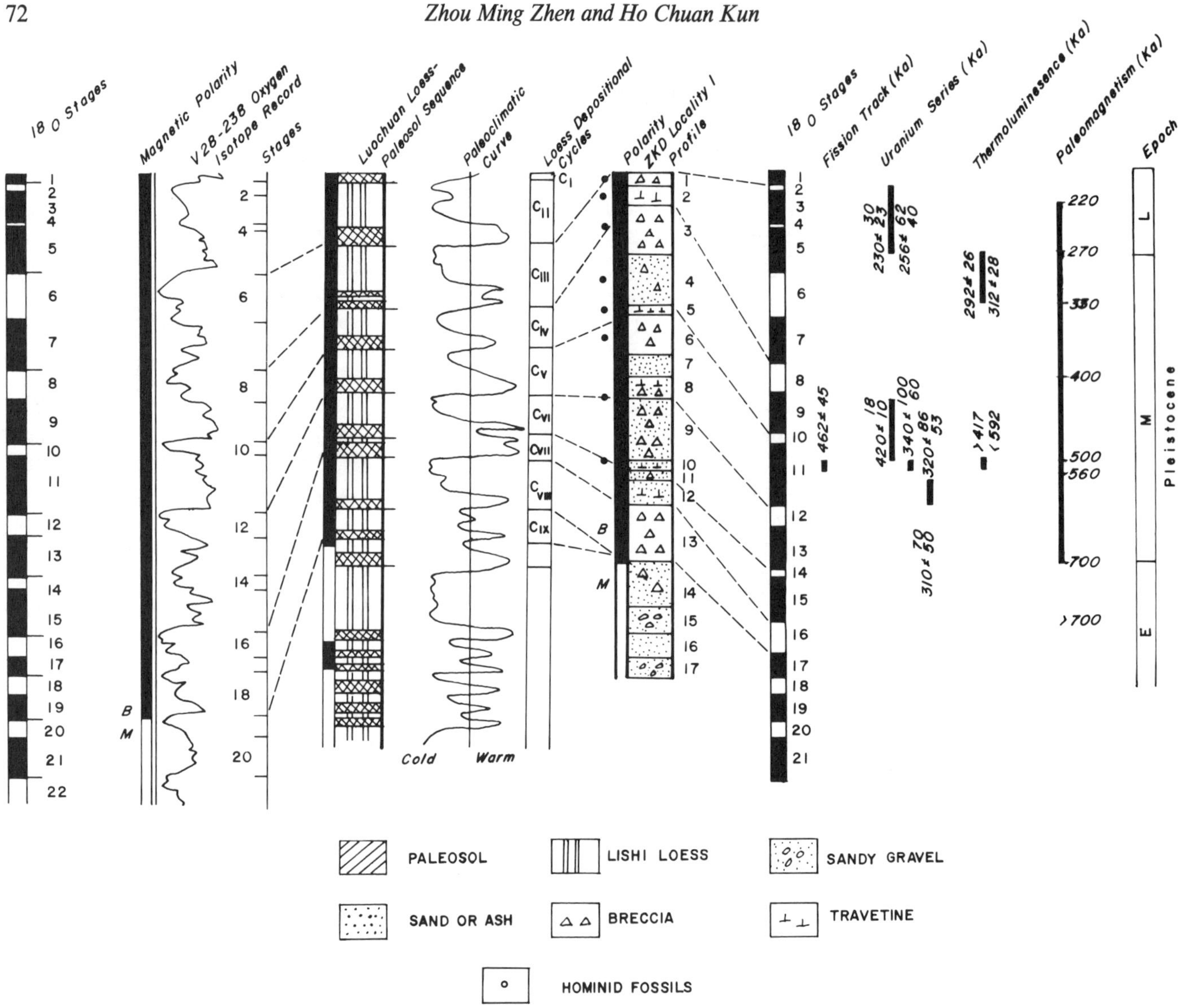

Figure 2. Correlations of absolute dates from Zhoukoudian Locality 1 with deep-sea cores and loess stratigraphic sequence of north China (see text for references).

cranial, mandibular, and dental samples from three time spans (early, middle, and late Pleistocene).

It is evident that under the climate of opinion in the late 1930s, students of human evolution tended to see in the evidence what they expected to see in light of their own theories. As late as the 1930s, theories of human evolution still were dominated by progressionism (Foley, 1989). Only later were they replaced by the renewed Darwinism of the modern evolutionary synthesis (Bowler, 1986).

During the 1950s and 1960s, better techniques of absolute dating confirmed that the time scale against which human evolution must be measured would have to be extended considerably. The length of the Pleistocene was doubled in 1961 from 1 to 2 m.y. (Evernden and Curtis, 1961). Equipped with more precise dating methods, both fossil hominid remains and their associations could be properly fitted into the new time scale. Logically, less effort should be concentrated on the "morphological dating" in terms of anatomical features, and more on the functional relations of morphology within paleoenvironmental contexts. Unfortunately, the Japanese invasion in 1937 and the civil war of 1948, as well as the Cultural Revolution, completely interrupted paleoanthropological research in China. Two themes that dominated the whole paleoanthropological field in the early 1950s and 1960s were the debate on the distinction of ape versus human and lithic/bone tools versus pseudotools.

With additional new discoveries of *Homo erectus* fossils of early Pleistocene age in north China (Lantian) we may even go one step further by discussing their adaptive radiation into different climatic zones (Lin, 1987). Recently, the hunting and fire-using ability of Zhoukoudian hominids has been questioned from a taphonomical perspective (Binford and Ho, 1985; James, 1989; see criticism Jia, 1989). In retrospect, without the geochronologic framework that has been laid out by foreign and Chinese geologists and paleontologists, which we have discussed above, it is impossible that the above-mentioned questions and past adaptations of Zhoukoudian hominids can be understood. Most of the geologic features of Zhoukoudian caves have been explained strictly in terms of "substantive uniformitarianism," as dubbed by Gould (1965), while environmental reconstructions and the lifeways of the cave occupants were based on "methodological uniformitarianism" (Gould, 1965). Even though we have finer chronological control than in the early days of the Zhoukoudian researches, the social organization of *Homo erectus* is still depicted as "more likely to have been living in a group than in solitude when he began to hunt deer. Furthermore, thousands of fossils of prey species found in the cave suggest that these primitive hunters may have preferred to bring prey back to the cave and share the meat with others in the community rather than to consume it where it was killed" (Wu and Lin, 1983, p. 94; Lin, 1985). On the contrary, after a closer look at Zhoukoudian faunas, Binford and Stone conclude that "previous interpretations presenting *Homo erectus* as a big-game hunter are no longer defensible. The use of fire may well have characterized the later occupations at Zhoukoudian, but none of these glimpses of the past justify the assumption that Zhoukoudian occupations were organized like modern man" (1986, p. 468). It is ironic that we still do not know very much about the behavioral adaptation of Zhoukoudian *Homo erectus* even after painstaking works of foreign and Chinese geologists, paleontologists, and paleoanthropologists at the foot of Dragon bone hill before 1977 and the first Multidisciplinary Project of Zhoukoudian after 1977.

ACKNOWLEDGMENTS

We appreciate the perceptive and critical comments received from two anonymous reviewers and Dr. Glover Krantz. Our special thanks are due to the editor, Dr. Laporte, for his encouragement, patience, and editorial assistance. The senior author thanks the K. C. Wang Foundation for a grant that helped defray research costs. Finally, we would like to dedicate this paper to Professor Jia Lan Po for his 80th birthday.

REFERENCES CITED

Aigner, J. S., 1986, The age of Zhoukoudian locality 1; The newly proposed ^{18}O correspondences; Fossil man; New facts, new ideas: Anthropos (Brno), v. 23, p. 157–174.

——, 1987, ^{18}O correlations and Zhoukoudian locality 1: L'Anthropologie, v. 91, p. 733–748.

An, Z. S., and Ho, C. K., 1989, New magnetostratigraphic dates of Lantian *Homo erectus:* Quaternary Research (in press).

Andersson, J. G., 1923, Essays on the Cenozoic of northern China: Geological Survey of China Memoir 3, series, A, p. 1–152.

Bada, J. L., 1987, Palaeoanthropological applications of amino acid racemization dating of fossil bones and teeth: Anthropologischer Anzeiger, v. 45, p. 1–8.

Binford, L. R., and Ho, C. K., 1985, Taphonomy at a distance; Zhoukoudian, the cave home of Beijing man?: Current Anthropology, v. 26, p. 413–442.

Binford, L. R., and Stone, N. M., 1986, Zhoukoudian; A closer look: Current Anthropology, v. 27, p. 453–475.

Black, D., 1926, Tertiary man in Asia; The Choukoutien discovery, *in* Leakey, L.S.B., Prost, J., and Prost, S., eds., Adam or ape?: Cambridge, Schenkman Publishing Company, Inc., p. 219–220.

Black, D., Teilhard de Chardin, P., Young, C. C., and Pei, W. C., 1933, Fossil man in China: Geological Survey of China Memoir 11, series A.

Bohlin, B., 1927, Excavation of the Chou Kou Tien deposit: Geological Society of China Bulletin, v. 6, p. 345–346.

Bowler, P. J., 1986, Theories of human evolution; A century of debate, 1844–1944: Baltimore, Maryland, The Johns Hopkins University Press, 318 p.

Chaney, R. W., 1935, The occurrence of *Celtis barbouri* at Choukoutien: Geological Society of China Bulletin, v. 14, p. 99–113.

Chaney, R. W., and Daughterty, L. H., 1933, The occurrence of *Celtis* associated with the remains of *Sinanthropus:* Geological Society of China Bulletin, v. 12, p. 323–328.

Chen, T. M., and Yuan, S. X., 1988, Uranium-series dating of bones and teeth from Chinese Palaeolithic sites: Archaeometry, v. 30, p. 59–76.

Derbyshire, E., 1983, On the morphology, sediments, and origin of the Loess Plateau of central China, *in* Gardner, R., and Scoging, H., eds., Megageomorphology: Oxford, Clarendon Press, p. 172–194.

Evernden, J. F., and Curtis, G. H., 1961, Age of Bed 1, Olduvai Gorge, Tanganyika: Nature, v. 191, p. 478–479.

Foley, R. A., 1989, In search for early man: Archaeology, v. 42, p. 27–32.

Gould, S. J., 1965, Is uniformitarianism necessary?: American Journal of Science, v. 263, p. 223–228.

Hsu (Xu), R., 1966, The climatic conditions in north China during the time of *Sinanthropus:* Scientia Sinica, v. 15, p. 410–414.

Huang, W. P., 1960, Restudy of CKT *Sinanthopus* deposits: Paleovertebrata et Paleoanthropologia, v. 2, p. 83–95.

James, S. R., 1989, Hominid use of fire in the lower and middle Pleistocene: Current Anthropology, v. 30, p. 1–26.

Jia (Chia), L. P., 1950, *Sinanthropus* (Peking man): Beijing, Longmen Union Press.

Jia, L. P., 1978, A note on the weather conditions in Choukoutien areas of Peking Man's time: Acta Stratigraphica Sinica, v. 2, p. 53–55.

——, 1989, On problems of the Beijing Man site; A critique of new interpretations: Current Anthropology, v. 30, p. 200–205.

Kahlke, H. D., and Chow, B. S., 1969, A summary of stratigraphical and paleontological observations in the lower layer of Choukoutien, Locality 1, and on the chronological position of the site: Vertebrata PalAsiatica, v. 3, p. 212–240.

Kahlke, H. D., and Hu, C. K., 1957, On the distribution of *Megaloceros* in China: Vertebrata PalAsiatica, v. 1, p. 273–284.

Kong, Z. C., and 9 others, 1985, Vegetational and climatic changes since Paleogene at Zhoukoudian and its adjacent regions, *in* Wu, R. K., and 7 others, eds., Multidisciplinary study of the Peking Man site at Zhoukoudian: Beijing, Science Press, p. 119–154.

Kurten, B., 1960, New evidence of the age of Peking Man: Vertebrata PalAsiatica, v. 2, p. 167–268.

Lewin, R., 1987, Bones of contention: New York, Simon and Schuster, 348 p.

Li, C., 1927, The Chou Kou Tien fossil deposits: Geological Society of China Bulletin, v. 6, p. 337–344.

Lin, S. L., 1985, Large fossil mammals of Locality 1 of Zhoukoudian and the hunting behavior of Peking Man, *in* Wu, R. K., and 7 others, eds., Multidisciplinary study of the Peking Man site at Zhoukoudian: Beijing, Science

Press, p. 95–101.
—— , 1987, Trends of distribution of early man in China: Acta Anthropologica Sinica, v. 6, p. 190–195.
Liu, C., Zhu, X. Y., and Ye, S. J., 1977, A paleomagnetic study on the cave deposits of Zhoukoudian (Choukoutien), the location of *Sinanthropus:* Scientia Geologica Sinica, v. 1, p. 26–33.
Liu, T. S., and Ding, M. L., 1984, A tentative chronological correlation of early human fossil horizons in China with the loess–deep sea records: Acta Anthropologica Sinica, v. 2, p. 93–101.
Liu, S. S., and 9 others, 1985, Study of the chronology of Peking Man site, *in* Wu, R. K., and 7 others, eds., Multidisciplinary study of the Peking Man site at Zhoukoudian: Beijing, Science Press, p. 239–240.
Liu, Z. C., 1985, Sequence of sediments of Locality 1 in Zhoukoudian and correlation with loess stratigraphy in northern China and with the chronology of deep-sea core: Quaternary Research, v. 23, p. 139–153.
—— , 1988, Paleoclimatic changes as indicated by the Quaternary karstic cave deposits in China: Geoarchaeology, v. 3, p. 103–115.
Movius, H., 1948, The lower Palaeolithic cultures of southern and eastern Asia: Transactions of the American Philosophical Society, v. 38, p. 330–420.
Oakley, K. P., 1966, Frameworks for dating early man: Chicago, Illinois, Aldine, 355 p.
Pei, J. X., 1985, Thermoluminescence dating of Peking Man site and other caves, *in* Wu, R. K., and 7 others, eds., Multidisciplinary study of the Peking Man site at Zhoukoudian: Beijing, Science Press, p. 256–260.
Pei, J. X., and Sun, J. Z., 1979, Thermoluminescence ages of quartz in ash materials from *Homo erectus pekinensis* site and its geological implication: Kexue Tongbao, v. 24, p. 849.
Pei, W. Z., 1939, An attempted correlation of Quaternary geology, paleontology, and prehistory in Europe and China: Geochronological Table, v. 1, p. 3–16.
Pei, W. Z., and Zhang, S. S., 1985, A study on the lithic artifacts of *Sinanthropus:* Beijing, Science Press, p. 1–277.
Pope, G., 1988, Recent advances in Far Eastern palaeoanthropology: Annual Review of Anthropology, v. 17, p. 43–77.
Qian, F., Zhang, J. X., and Yin, W. D., 1979, Magnetic stratigraphy from the sediments of West Wall and test pit at Zhoukoudian; Conference Commemorating the 50th Anniversary of the Discovery of the First *Sinanthropus:* Beijing, Institute of Paleontology and Palaeoanthropology, p. 34–37.
—— , 1985, Magnetic stratigraphy from the sediment of West Wall and test pit of Locality 1 at Zhoukoudian, *in* Wu, R. K., and 7 others, eds., Multidisciplinary study of the Peking Man site at Zhoukoudian: Beijing, Science Press, p. 151–255.
Sun, M. J., 1965, On the pollen from the Zhoukoudian: Quaternalia Sinica, v. 4, p. 84–96.
Teilhard de Chardin, P., 1936, Fossil mammals from Locality 9 of Choukoutien: Paleontologica Sinica, series c, v. 7, fasical 4.
Teilhard de Chardin, P., and Young, C. C., 1929, Preliminary report on the Chou Kou Tien fossiliferous deposits: Geological Society of China Bulletin, v. 8, p. 173–202.
Von Koenigswald, G.H.R., 1983, Davidson Black, Peking Man, and the Chinese dragon, *in* Sigman, B. A., and Cybulski, J. S., eds., *Homo erectus;* Papers in honor of Davidson Black: Toronto, Ontario, University of Toronto Press, p. 27–39.
Weidenreich, F., 1936, *Sinanthropus pekinensis;* A distinct primitive hominid: Proceedings of the Joint Meeting of the Anthropological Society of Tokyo and the Japanese Society of Ethnology, 1st session, p. 13–37.
Wolpoff, M. H., 1984, Evolution in *Homo erectus;* The question of stasis: Paleobiology, v. 10, p. 289–406.
Wu, R. K., and Lin, S. L., 1984, Peking Man: Scientific American, v. 248, p. 86–94.
Wu, X. Z., and Wang, L. H., 1985, Chronology in Chinese paleoanthropology, *in* Wu, R. K., and Olsen, J. W., eds., Palaeoanthropology and palaeolithic archaeology in the People's Republic of China: New York, Academic Press, p. 29–51.
Xu, Q. Q., and Ou, Y. L., 1982, Climatic changes during Peking Man's time: Acta Anthropologica Sinica, v. 1, p. 80–90.
Xu, Q. Q., and You, Y. Z., 1982, Four post-Nihewanian Pleistocene mammalian fauna of north China; Correlation with deep-sea sediments: Acta Anthropologica Sinica, v. 1, p. 180–190.
Yang, Z. G.,a nd 7 others, 1985, Study of the Late Cenozoic strata at Zhoukoudian, *in* Wu, R. K., and 7 others, eds., Multidisciplinary study of the Peking Man site at Zhoukoudian: Beijing, Science Press, p. 1–85.
Yih, L. F., 1920, The geology of the Hsi-shan or the Western Hills of Peking: Geological Survey of China Memoir 1, series A, 115 p.
Zdansky, O., 1923, Uber ein Saugerknochenlager in Chou Kou Tien: Geological Survey of China Bulletin, v. 5, p. 83–89.
Zhou, B. X., 1979, The fossil rhinocerotids of Locality 1, Choukoutien: Vertebrata Palasiatica, v. 1, p. 29–50.
Zhou, M. Z., 1955, Living natural environments of Chinese fossil men in terms of vertebrate mammalian fossils, *in* Kuo, M. J., and 5 others, eds., The discovery and research of Chinese human fossils: Beijing, Science Press, p. 19–38.
—— , 1981, Vertebrate paleontology in China, 1949–1979, *in* Teichert, C., Liu, L., and Chen, P. J., eds., Paleontology in China, 1979: Geological Society of America Special Paper 187, p. 15–20.
Zhao, S. S., and 6 others, 1985, Uranium-series dating of Peking Man site, *in* Wu, R. K., and 7 others, eds., Multidisciplinary study of the Peking Man site at Zhoukoudian: Beijing, Science Press, p. 246–250.

MANUSCRIPT ACCEPTED BY THE SOCIETY JUNE 6, 1989

Typeset by WESType Publishing Services, Inc., Boulder, Colorado
Printed in U.S.A. by Malloy Lithographing, Inc., Ann Arbor, Michigan